WERNHER VON BRAUN

OFFICE OF MILITARY GOVERNMENT FOR GERMANY (U. S.)
APO 742

FILE NO: DC/20779/158 T
SUBJECT: NSDAP Records Check

NO.	TO	FROM	DATE	(Has this been coordinated with all concerned?)
1	S-2 BC	7771 Doc. Ctr.	23 Apr 47	Att'n.: Capt Hirsch

1. Reference telephone conversation this date.

2. The following information is certified as being a true extract from original Nazi party records in the custody of this Center:

Dr. phil. Wernher von BRAUN
Born: 23 Mar 1912 at Wirsitz
Occupation: Technischer Leiter & Professor d. Heeresversuchsanstalt Peenemuende
Party-No.: 5738692
Entered Party: 1 May 1937
SS-No.: 185068 ("Wiederaufnahme"1 May 1940)
Untersturmfuehrer 1 May 1940
Obersturmfuehrer 9 Nov 1941
Hptsturmf. 9 Nov 1942
Sturmbannf. 28 Jun 1943
Assigned: Stab Oberabschnitt Ostsee
8 Semesters Technische Hochschule & University majored in Techn. Physik (studied in Switzerland: March/Sept 1931).
Military Service (Luftwaffe) 1.5.36 - 15.6.38
Flugzeugfuehrerschule Frankfurt/Oder & Stolp.

FOR THE COMMANDING OFFICER:

KURT ROSANOW
US- Civilian
Chief of Branch

Tel.: 44 344

16 8

Wernher von Braun's membership in the Nazi Party, ranks held in the SS, and service in the Luftwaffe are summarized in a document from the Office of Military Government for Germany (U.S.). From Army FOI file on von Braun, 168.

WERNHER VON BRAUN

The Man Who Sold the Moon

Dennis Piszkiewicz

Westport, Connecticut
London

Library of Congress Cataloging-in-Publication Data

Piszkiewicz, Dennis.
    Wernher Von Braun : the man who sold the moon / Dennis
Piszkiewicz.
        p.   cm.
    Includes bibliographical references and index.
    ISBN 0-275-96217-2 (alk. paper)
    1. Von Braun, Wernher, 1912-1977. 2. Rocketry—United States—
Biography. 3. Rocketry—Germany—Biography. I. Title.
TL751.85.V6P57   1998
621.43'56'092—dc21
[B]   98-14481

British Library Cataloguing in Publication Data is available.

Library of Congress Catalog Card Number: 98-14481
ISBN: 0-275-96217-2

First published in 1998

Praeger Publishers, 88 Post Road West, Westport, CT 06881
An imprint of Greenwood Publishing Group, Inc.

Printed in the United States of America

The paper used in this book complies with the
Permanent Paper Standard issued by the National
Information Standards Organization (Z39.48-1984).

10 9 8 7 6 5 4 3 2 1

For my parents,
William and Harriette Piszkiewicz

Contents

Photographs follow pages 81 and 150

Preface

Wernher von Braun played the role of scientist as celebrity better than anyone of his era, including Albert Einstein. His story has been told many times by his friends, colleagues, admirers, and by himself. These accounts have always presented his life as he would have told it, yet he kept a closet full of secrets about his early career in Nazi Germany that would have horrified his admirers and those whose approval he courted.

During his sixty-five years, von Braun traveled through the technological, political, and moral terrain of the twentieth century on his journey to outer space. Most Americans knew that he designed the V-2 missile for Hitler's Germany, and most accepted his explanation that he was young and naive, that he worked with the Nazis in order to pursue his real interest, the development of rockets for the exploration of space, but that he had never believed in the Nazi cause. After the destruction of the Third Reich, von Braun came to America, where he worked first for the U.S. Army developing the missiles that were an essential armament of the Cold War, and later for NASA building the huge rockets that would take America into space and carry the Apollo astronauts to the moon.

Wernher von Braun presented an image of himself that many still remember: the aristocratic good looks, the smooth German accent, the authoritative air, the unashamed belief in science and technology, and the certainty that space travel was not only possible but inevitable. He appeared on television, before Congress, and at college commencements with his message. He wrote a seemingly endless torrent of articles and books that all flowed in the direction of space. He worked for the United States as an engineer and energetic manager, but he worked for his cause with even more dedication. He was the prophet of the space age; he was the man who sold the moon.

Wernher von Braun: The Man Who Sold the Moon tells of von Braun's work for the Army and NASA and of his three decades of tireless promotion of space travel. But it also examines his Nazi past, which he tried so hard to conceal. This book recounts von Braun's early years in Germany developing the V-2, his memberships in the Nazi party and the SS, his complicity in the Nazi cause and its crimes, the cover-up of his past when he immigrated to the United States, and the inevitable unraveling of the myth Wernher von Braun had so carefully created about himself.

Acknowledgments

Many fine institutions helped me to discover the story of Wernher von Braun. I wish to thank the following: the Orange County Public Library, the Los Angeles County Public Library, the Simon Wiesenthal Center in Los Angeles, the library of the California State University at Fullerton, the library of the University of California at Irvine, and the Library of Congress. I also thank the staffs of the National Archives in Washington, D.C., NASA Headquarters in Washington, D.C., the Redstone Arsenal, the Marshall Space Flight Center, and the Freedom of Information Offices of the United States Army and of the FBI.

I thank the following for their help in locating and making available historic photographs: the Marshall Space Flight Center, the Redstone Arsenal, NASA Headquarters in Washington, the National Archives, and the National Air and Space Museum of the Smithsonian Institution.

Many people helped me bring this project to its conclusion. I especially wish to thank Aaron Breitbart of the Simon Wiesenthal Center in Los Angeles for sharing his files on Wernher von Braun and his rocket team. I am indebted to the Kaylor clan, Marvin, Wilma, and, as always, P. J. for their help in the form of suggestions, criticisms, and proofreading. I also thank Eric Knabe for reviewing the manuscript and giving me his perspective on history.

WERNHER VON BRAUN

Prelude: The Rocket

The long-studied assault on England by unmanned missiles now began: the target was Greater London.

The Rocket was an impressive technical achievement. . . . Its maximum speed was about four thousand miles an hour, and the whole flight took no more than three or four minutes.

Winston S. Churchill[1]

The cortege that made up the first battery of the Wehrmacht 485th (Mobile) Artillery Detachment rolled toward a crossroads on the outskirts of the Hague bearing its burden of creation and destruction, triumph and terror, dreams and death.[2,3] The parade was led by three half-tracks towing *Meillerwagens*; each half-track carried a firing crew, and each *Meillerwagen* a rocket that was over forty-five feet long with a diameter of over five feet. They were followed by three tank trucks, one with alcohol fuel, another with liquid oxygen, and the third with auxiliary fuels and equipment. These were followed in turn by a truck carrying an electric power generator, then a truck loaded with guidance equipment. Staff cars with officers brought up the rear.[3] The second battery of the 485th (Mobile) Artillery Detachment also moved into position at a crossroads nearby. Although each battery was capable of firing three rockets, this day they would prepare only one each.

The firing crew placed a firing table on the pavement, then raised a rocket onto it, orienting it so that when launched, it would follow a trajectory to the west. The tank trucks pulled into position to begin filling the rocket with its fuels. Several hours later, the trucks and personnel had pulled back to a safe distance, just in case something unexpected happened. The only connection between the missile and its crew were the cables that snaked from it to the trucks that carried the electric power generator and the guidance equipment. A broad band of white frost, condensed moisture from the air, circled the lower part of the rocket at the level of its liquid oxygen tank. Near the stern of the rocket, oxygen vapor escaping through a vent cooled moisture from the air into a small, billowing cloud. Then the cloud dissipated as the firing crew closed the oxygen vent valve.

A hail of sparks shot down through a cloud at the bottom of the rocket, bouncing off the cone-shaped blast deflector and scattering across the pavement. The two umbilical cables that tied the missile to earth fell away. The sparks from the rocket motor gathered into an orange flame, then grew in power to twenty-five tons of savage thrust. At 6:38 P.M. on September 8, 1944, the V-2 rose from the firing table, quickly gained speed, and rushed into history. The rocket and its thunder faded into the distance, in the direction of the setting sun, leaving behind a trail of condensed exhaust vapor.[4] Within seconds of the launch of the first missile, a second lifted off from a nearby crossroads on the outskirts of the Hague.

That day in London, 200 miles to the west, a sense of safety, if not peace, descended on the beleaguered British capital. The Allied Forces had invaded the continent four months earlier, and with that success, they destroyed or captured the launching ramps the Luftwaffe used to fire V-1 cruise missiles toward London.[5] The city was free of the pilotless missiles that brought destruction to it on their dead-reckoning flights. On the evening of September 8, 1944, Londoners were contentedly on their way home from work or preparing an evening meal.

Fifty-eight seconds after liftoff, according to plan, the firing crew sent a radio signal to the missile that cut off the flow of fuel to the rocket motor, which brought about *Brennschluss*, the end of burning. The missile, now traveling at almost 3,500 miles per hour continued rising, coasting without power into the west on a ballistic arch. It followed a path governed by the laws of physics, of inertia and gravity.

At 6:43 P.M. without warning, a ton of high explosives detonated in Chiswick, a half-dozen miles west of the center of London. The blast tore a thirty-foot-wide crater in the center of the concrete roadway, demolished six surrounding houses, and heavily damaged at least twice as many more. As the rubble thrown up by the blast was still falling, London heard a loud, double boom. The report, which some thought was like a crack of thunder, was the twin shock waves caused by the V-2 missile smashing through the sound barrier as it slammed back into the atmosphere on the downward leg of its journey, arriving after the missile's lethal cargo.

When emergency crews arrived at Chiswick, they found that the blast impersonally snuffed out three lives and severely injured ten more people. The rocket, incarnated as the modern ballistic missile, made its first kill in battle.

Sixteen seconds after the first rocket shattered Chiswick, the second tore into Epping, about fifteen miles northeast of central London. It destroyed a few wooden huts, but caused no serious damage or casualties.[2,3] The two missiles that struck England followed one that fell on Paris at 8:34 that morning, and was reported to have caused modest damage.[6]

For the next seven months, Nazi Germany, in its last desperate attempt to turn the course of the war with a campaign of terror bombardment, fired over 1,300 V-2 rockets at England. When the rubble was cleared and the bodies counted, the English tallied 518 V-2s that fell on London and 537 that struck elsewhere in England.[7] The missiles killed 2,724 people and seriously injured another 6,467.[8] Germany struck Antwerp with 1,265 V-2s, and hit Paris and other targets with hundreds more. An accurate count of casualties on the continent is not available.

In the context of the Holocaust that was World War II, the casualties resulting from the V-2 campaign—while unquestionably causing personal tragedies—were insignificant. The dead were simply the victims of faceless soldiers who sent anonymous death from the sky. Several years would pass before the world knew the name and face of the man who created the rocket and who had, in the process, changed the nature of war. Decades would go by before the same man would change the world's view of the universe.

I

The Immigrant

I always was a German and I still am.

Wernher von Braun, 10 June 1945[1]

The young man who stepped off the C54 cargo plane thought of himself as an immigrant who like many before him came to America to pursue his dream, build a new life, and leave behind crimes of the past. His arrival at Newcastle Army Air Base in Wilmington, Delaware, on September 18, 1945 was a second birth, and the official circumstance was a baptism that cleansed him of the original sin of his first life.[2]

He arrived not as a free man, however, but in the custody of the United States Army. He was one of a group of seven German rocket experts, the vanguard of a larger team that the Army recruited under Operation Overcast (later renamed Project Paperclip) to exploit for their technical expertise. He was the senior member of the group although not the oldest, and he had been its leader in Germany since the beginning, even before the Nazis came to power. His name was Wernher von Braun.[3]

The day after their arrival, the seven German men boarded a DC-3 for the relatively short hop to Boston's Logan Field.[4] On landing, they went by Army sedan to Boston Harbor, where they boarded a boat for the five-mile trip to Fort Strong on Long Island.

Fort Strong was a post of the United States Army Intelligence Service, and the Germans were there to be interrogated and processed. The questioning was a repeat of the process they went through months earlier in Germany after they surrendered to the Army. At the end of it, the Army compiled the "Basic Personnel Records" of their charges and new employees. Wernher von Braun's record indicated that he was thirty-four-years-old, six-foot one and one-half inches tall, weighed 172 pounds, with blue eyes, blond hair, and a fair complexion. The record noted as distinguishing marks an operation scar on his left lower arm and a scar on his upper lip.[5]

While von Braun was certainly happy to be in America, he was physically miserable. The previous winter, in the final days of World War II, he was in an automobile accident, which accounted for the scars cited in his personnel record. His shoulder was shattered, his left arm was broken in two places, and had not been set properly.[6] While he no longer wore the paralyzing cast, his arm still ached, and the pain was exacerbated by the foul weather. To add to his misery, he had contracted hepatitis, presumably before leaving Europe.[3]

To fill in their free time, the Germans played marathon games of Monopoly or took long walks in the fall air on the beaches of Long Island.[3] The dreary interlude at Fort Strong ended on October 1, 1945, when Maj. James P. Hamill signed documents accepting custody of the seven Germans. That same day, the War Department announced in a press release to the American people the presence in the United States of the German rocket scientists and others in Operation Overcast.

OUTSTANDING GERMAN SCIENTISTS BEING BROUGHT TO U.S.

The Secretary of War has approved a project whereby certain outstanding German scientists and technicians are being brought to this country to ensure that we take full advantage of those significant developments which are deemed vital to our national security.

Interrogation and examination of documents, equipment and facilities in the aggregate are but one means of exploiting German progress in science and technology. In order that this country may benefit fully from this resource a number of carefully selected scientists and technologists are being brought to the United States on a voluntary basis. These individuals have been chosen from those fields where German progress is of significant importance to us and in which these specialists have played a dominant role.

Through their *temporary stay* [emphasis added] in the United States these German scientists and technical experts will be under the supervision of the War Department but will be used for appropriate military projects of the Army and Navy.[7]

This announcement went virtually unnoticed in the American press. It was not until a year later when the Army revealed specific details and identified the German scientists that the American people reacted, both positively and negatively.

Major Hamill's assignment was to start the Army's first guided-missile program, and he now had his first seven workers. Hamill took von Braun to Washington, D.C., to meet with high-ranking Army Ordnance officers. He sent the other six Germans to the Aberdeen Proving Ground in Maryland. Their first assignment was to sort through and organize the seven tons of technical documents they hurriedly packed and carried with them from their operational headquarters at Peenemuende in an organized retreat from the advancing Soviet Army.[4]

After spending several days in Washington discussing rockets with Army brass, Hamill took von Braun to Union Station, where they boarded a train for the first leg of their journey to El Paso, Texas, via St. Louis. At St. Louis, Hamill learned that their tickets for the next leg placed them on a train that had as passengers only wounded veterans of the 101st and 82d Airborne Divisions. He thought it unwise to chaperone von Braun among men who had fought against Germany and suffered, and, with difficulty, he was able to get seats on a standard passenger train.

When Hamill and von Braun boarded the train, they found it was full and that their section assignments were at opposite ends of the same car. Hamill was not happy with the situation since his orders were to keep von Braun under his control and not allow him to talk to anyone. However, since von Braun was still within sight, Hamill accepted the arrangement as the fastest way of getting to their destination.

A man sharing the same section of the Pullman car with von Braun struck up a conversation, and the gregarious von Braun joined in. The man, noticing von Braun's unmistakable accent, asked where he was from.

Switzerland, von Braun lied.

The man said he knew Switzerland quite well. He asked von Braun what business he was in.

Von Braun answered vaguely that he was in the steel business.

To von Braun's distress, the man had also been in the steel business, and he pressed von Braun for more information about his life and work in Switzerland. Von Braun improvised with what little knowledge he had picked up as a student while residing in Zurich for half-a-year, and as a tourist.

The man asked von Braun about the segment of the steel business he worked in.

Ball bearings, von Braun answered.

The man was also familiar with the ball bearing business.

Fortunately for von Braun and Hamill, the train arrived at Texarkana, the man's stop. As the man departed, he warmly shook von Braun's hand and said, "If it wasn't for you Swiss, I doubt if we could have beaten those Germans."[8]

Major Hamill and Wernher von Braun arrived at El Paso, Texas, on October 3, 1945. Their base was situated in El Paso, at Fort Bliss.[2] Von Braun spent the night lodged at the junior officers' Bachelor Officers' Quarters, where Hamill also stayed so he could formally keep von Braun in his custody. The following morning, von Braun was clearly suffering from a bout of hepatitis, and he was admitted to the William Beaumont Army Hospital. Major Hamill transferred responsibility for custody of von Braun to the hospital surgeon, and was then free to go about his real job, that of setting up the Army's guided-missile program.[9]

The staff at the William Beaumont Army Hospital put von Braun on a fat free diet and prescribed several weeks of rest. His stay turned into a surprisingly pleasant interlude. He wrote:

> I had assumed I would encounter plenty of hostility as an "ex-enemy big shot." But I never did. In America you don't seem to carry grudges, as do many Europeans who have been enemies. . . . Officials advised me to hide my identity, but I couldn't conceal my broken English. The GI's sized me up with uncomfortable accuracy, and began calling me "The Dutchman." But they also invited me to join their blackjack poker games![10]

The city of El Paso is at the westernmost tip of Texas, where the Rio Grande winds up from between Texas and Old Mexico to meet the state of New Mexico. The El Paso area boasts the oldest settlement in Texas, the Mission Nuestra Señora del Carmen, at Ysleta, founded in 1682 to serve Spanish immigrants and Native Americans. Texas became part of the United States with its annexation in 1845, and the Army established Fort Bliss at El Paso in 1848. When Wernher von Braun arrived a century later, El Paso was a small urban outpost in the middle of the sparsely populated, undeveloped American West.

Wernher von Braun found the landscape less welcoming than its people. He described the desert and his reaction on first seeing it:

> One evening [I was] . . . watching the sun, beyond the tremendous expanse of desert, set on the Sacramento Mountains. The range was well over a hundred miles distant, but in the clear desert air it looked as if it were but a stone's throw away. To my continental eyes, accustomed to the verdant green valleys and hills of central Europe, the sight was overwhelming and grandiose, but at the same time I felt in

my heart that I would find it very difficult ever to develop a genuine emotional attachment to such a merciless landscape which, while unable to support more than a mere trace of vegetation, dwarfed man by its very expanse.[11]

Although the Army selected Fort Bliss as the center for its rocket research program, it also needed a testing range. It elected to fire its rockets at the new White Sands Proving Ground in the Tularosa Valley of New Mexico. The rocket launch site was about thirty-five miles north of El Paso as the crow flies, and about fifty miles by road. White Sands was remote, yet it was surrounded by important historical sites. About eighty-five miles due north and down range from the White Sands launch site was Trinity Site, the remote point in the desert where, the previous July, J. Robert Oppenheimer and his group of nuclear physicists detonated the first atomic bomb. To the northeast, 120 miles from the launch site, was the Mescalero Ranch near Roswell, where American rocket pioneer Robert H. Goddard launched his rockets in the 1930s.[12]

While von Braun was in the hospital recovering from his attack of hepatitis, Major Hamill prepared for the arrival of his German employees. Hamill went out to White Sands to see the preparations that had been made there. He found two shacks and not much else.[8] It was not the modern technical center the Germans had at Peenemuende, but then Peenemuende was no longer what it had been, not after the RAF bombed it in August 1943[13] and the Germans stripped it of anything worth taking in early 1945.[14] In a few more days, Hamill identified the Walter Beaumont General Hospital Annex as an ideal residence for the German rocket scientists. It was a group of pleasant barracks-style buildings within a security perimeter. After some negotiations, it was transferred to his command.[9]

On February 23, 1946, the last of the 118 German rocket scientists brought to the United States under the umbrella of Operation Overcast arrived at El Paso. (Regrettably, one of the members died soon after his arrival, reducing the size of the group to 117.) They were identified as Department of the Army Special Employees (DASE).[15] Most of the German DASEs were quartered at Fort Bliss, although a group of twelve to fifteen was sent the short distance north to White Sands, where they were to assemble rockets.[16] *The Team*, as they were called, was again intact and ready to function in a new country.[17] It would do so as an insular group, both respected and resented, until the first men—native-born Americans—walked on the moon.

Although the group of German rocket scientists had no formal name, it was known by its members, friends, and detractors in the years that followed as "The Team," "The Rocket Team," "The Peenemuende Team," "The Germans," and "von Braun's Team." It had no formal organization outside of Operation Overcast and the various government organizations that succeeded it. The team, however, had strict de facto membership rules and a feudal organization. All team members were German-born. They had all been part of von Braun's German Army rocket development group, many since its formation in the early 1930s. They were all loyal to von Braun. Their egos let them live happily in the shadow of the team's leader and spokesman, Wernher von Braun.

The team began in October 1932 when its first member and leader, twenty-one-year-old Wernher von Braun, went to work for the German Army's ballistics and ammunitions group to develop liquid-fueled rockets as weapons. The Army officer who headed the group was Colonel, later General, Walter Dornberger.[18] In the thirteen years that followed, several thousand scientists and engineers joined the group, which had as its technical director the young von Braun. By the end of World War II, the team was dispersed and its technical center at Peenemuende destroyed; however, a small group, led by von Braun surrendered to the United States Army and offered their services to the victors. The Army took them up on the offer, and gave the 118 men six-month contracts.[19]

Since the United States Army had little expertise with rockets and none with big missiles such as the V-2, it went to the leader of the group to identify the key technical experts of his team who could effect the transfer of technology. Von Braun's selection of members of his team showed an adherence to his priorities and loyalties as well as to the Army's needs. Among the 118 members of the Team who came to Fort Bliss were:

- *Walter Riedel*, an engineer who was involved in German amateur rocketry in the late 1920s and early 1930s. Riedel was one of the first men who joined von Braun to work on rockets for the German Army.[18]

- *Arthur Rudolph*, also a participant in German amateur rocketry and an early recruit to von Braun's German Army Team.[20] Rudolph had been chief operations director of the Mittelwerk factory in Germany that had built the V-2s.

- *Herbert Axster*, a lawyer in civilian life and a lieutenant colonel in the defeated German Army.[21] Axster had been chief of staff to General Walter Dornberger, who led the German Army's rocket development program for thirteen years. Axster was a relative newcomer, having joined the Peenemuende group late in World War II.[22]

- *Magnus von Braun*, Wernher von Braun's younger brother, who was trained in chemistry but had been in charge of gyroscope production for the V-2s.[23]

The Germans who came to the United States on temporary contracts under Operation Overcast left their families and loved ones behind. Their dependent wives, children, and parents, were housed in a camp operated by the United States Army in Landshut in Bavaria. The standard of living in the compound was better than that of the surrounding area, and the locals who were excluded resented the clear discrimination. Not surprisingly, Operation Overcast was an open secret, and the residence for dependents was commonly known as "Camp Overcast." Because of this embarrassing breach of secrecy and because of continuing expansion and formalization of the program, Operation Overcast became Project Paperclip on March 13, 1946.[24]

While the German scientists may have had some difficulty adjusting to the geography, weather, and food in the American desert, they were initially inclined to work for the United States Army. It was nothing new. In Germany, most of them had been civilians engaged in rocket development for its army. They had been following orders for years and were prepared to continue doing so in the future. Only their master had changed.[25]

As defined by the Chief of the Technical Division, Office of the Chief of Ordnance of the Army, their responsibilities were threefold. They were to train military, industrial, and academic personnel in the design, construction, and operation of rockets and guided missiles. They were to assemble V-2 rockets from parts that were shipped to White Sands from the Mittelwerk factory in central Germany and launch them from White Sands. They were also to investigate the military and research applications of rockets. The Army contracted with the General Electric Company to give technical support and to ensure that the United States got the most out of its new employees.[26]

Although American personnel were to retain control of the entire operation, the Germans were to test fire the rockets, since they knew how the rockets functioned. In a static test on March 15, 1946, their first rocket burned for fifty-seven seconds until its fuel was exhausted. A month later on April 16, a second V-2 was launched, however, it's motor was cut off by an emergency radio signal after nineteen seconds when a stabilizing fin tore free.[27,28] The rocket reached an altitude of five miles. The first successful flight of a V-2 from White Sands came on May 10, 1946. The missile reached its programmed altitude of seventy miles and covered a range of thirty-one miles.[27]

The fourth V-2 launched from White Sands on May 29, 1946, demonstrated the missile's potential for political as well as physical damage.[28] The missile ascended as planned, but instead of arcing over to the north, to the far end of the firing range, it leaned to the south. The gyroscopic guidance system had obviously malfunctioned. Standard procedure for such a failure was for the range safety officer to prematurely cut off the missile's fuel supply by radio signal. The safety officer was Ernst Steinhoff, a member of von Braun's German team, who was to give the fuel cut off order to a sailor from the Naval Research Laboratories, who handled the signal-transmission equipment. Steinhoff understood that if the rocket's flight ended prematurely, it would fall to earth with a good part of its load of explosive fuel splattered over the landscape south of White Sands, the city of El Paso, or Ciudad Juarez, even farther south beyond the Rio Grande. Steinhoff decided to let the rocket burn its fuel and fall back to earth harmlessly in the unpopulated area south of Juarez. It did, barely.

The V-2 smashed into a rocky hillside adjacent to Tepayec Cemetery, one and one-half miles south of Juarez. The burnt-out rocket carried with it enough kinetic energy to blast a hole thirty feet deep and fifty feet in diameter.

Lt. Col. Harold R. Turner, commander of the White Sands Proving Ground, immediately began damage control. He contacted the commanding general of the Mexican state of Chihuahua, and he learned that there were no casualties and no damage beyond the hole in the ground. The Mexican general would take care of any diplomatic problems with the authorities in Mexico City. Washington was less accommodating. Turner soon found himself explaining the incident to Gen. Dwight D. Eisenhower, chief of staff of the United States Army, and Secretary of State George C. Marshall. Later, a board of investigation did a formal analysis of the incident and supported Steinhoff's decision.

Men of lesser responsibility had a simpler perspective of the incident. One lieutenant from White Sands boasted that he was part of the first United States Army unit to fire a guided missile against a foreign country. Across the Rio Grande, Mexican entrepreneurs were making a buck (or peso) from the V-2 even before its wreckage cooled. Concession stands were reportedly selling unidentifiable metal fragments as souvenirs of the V-2 within ten minutes of its crash. Ernst Steinhoff estimated dryly that the Mexican vendors sold at least ten to fifteen tons of fragments (some of them suspiciously resembling fragments of tin cans) from the missile that had an empty weight of four tons.[29]

The whole purpose of Project Paperclip was, of course, to transfer technology from the defeated Germans to their new superiors, the United States. Besides the practical demonstrations of rocket assembly and launchings conducted at White Sands, the Germans organized documents and were generating reports at an impressive rate. By April 19, 1946, Maj. James Hamill had supervised completion of a dozen reports that he then circulated to Army groups from headquarters in Washington, D.C., to the Jet Propulsion Laboratory in Pasadena, California.[30] And if the Army did not have enough ideas of its own regarding rockets, Wernher von Braun suggested new projects.[31]

By mid-1946, many of the German rocket scientists and engineers had been in the United States for nearly a year. For security reasons and because they were formally, what would years later be called "undocumented aliens," their travel had been severely restricted by the Army. The group at Fort Bliss was restricted to their compound, an area of about six acres.

They could leave only with an Army escort.[32] In early September, the Army relaxed its restrictions and gave the German DASEs passes that allowed them to travel freely anywhere within the greater El Paso area. In late November, the Army distributed to the Germans for their signatures contracts that extended their employment into a second year.[33] The Army's initial intention was to acquire the technology of liquid-fueled rockets in general and an understanding of the V-2 in particular. As the first year stretched into the second, and then into the long years of the Cold War, the Army and the United States also kept the Germans as permanent employees.

The presence of the German rocket scientists at Fort Bliss and White Sands was an open secret from the beginning; but with their dependents on the way from Germany to join them—to double and possibly triple the size of their colony—there was no way to keep this secret from mutating into awkward and embarrassing public knowledge. The Army took the offensive by presenting their rocket experts to the media as law-abiding government employees and soldiers in the battle for peace and against Communism.

The news media went along. Soon, the people of El Paso, if not the general population of the United States, learned more about the immigrants than they probably wanted to know, although they did not learn anything without the Army's approval. The media blitz was a sustained exercise in calculated openness:

"German Scientists to Be Interviewed. German scientists of whom Fort Bliss officials speak in whispers will answer questions for reporters soon" (*El Paso Herald-Post*, November 13, 1946).[34] During the years of the Third Reich, Wernher von Braun and his work had been state secrets. The people of Germany learned about the V-2 rocket in the last months of the war, and nobody without a need to know ever heard of von Braun and his team. (When he surrendered to the United States Army at the end of the war, he was photographed and interviewed by reporters, but then quickly forgotten. On arrival in the United States, he once again became a state secret.) Von Braun was no doubt thrilled by the United States Army's openness because of his own desire for recognition and because he had an agenda. He would step into the spotlight like a band leader in front of his orchestra.

"118 Top German V-2 Experts Stationed in E.P.—Builders of Nazi Secret Weapons Working in U.S." (*El Paso Times*, December 4, 1946).[34] The same day, the *El Paso Herald-Post* ran a story with photographs about the refueling of a V-2 missile.

"German Scientists Tell How V-2 Rocket Was Developed from Experimental Mile High Missile" (*El Paso Herald-Post*, December 5, 1946). In this front-page article, Wernher von Braun explained to the press the development of the V-2 and his surrender to the United States Army along with the members of his team. He had little to say about the use of the missile during the war.[35]

"V-2 Tests 33% Successful" (*El Paso Times*, December 6, 1946). The body of the article corrected its title by stating that 33 percent of firings of the V-2s from White Sands were unsuccessful, the remaining 67 percent were, presumably, successful. Wernher von Braun spent three hours fielding questions from reporters about rockets, then he announced his imminent marriage to Maria Louise von Quistorp. The Germans were saying that the ceremony would be performed under "crossed rockets."[36]

"V-2 Rocket Sets New Speed Record." In the same issue, the *El Paso Times* reported on a successful firing in which a V-2 reached a speed of 5,000 feet per second (approximately 3,400 miles per hour). The short article also announced a rocket launch for December 17 that would release "man-made meteorites," fireworks for the people of Texas and New Mexico.[37]

"German Scientists Have Own Court to Handle Breaches of Contract" (*El Paso Herald-Post*, December 6, 1946).[38] By their presence in the United States without passports or visas, the German rocket scientists were breaking the law, even though they were sponsored by and in the custody of the Army. Many innocent things they might do, such as traveling to areas beyond the Army's control, could bring serious embarrassment to the Army, jeopardize their presence in the United States, and risk their hopes for careers building rockets. Major Hamill asked von Braun to set up a self-policing body within his group. This "people's court," as it became known among the Germans, judged infractions of good judgment, if not the law, and determined punishment. Von Braun appointed Dieter Huzel as judge, and he also named several others as jurors.[39]

A fascinating example of the kind of infraction the people's court dealt with and how it delivered justice is found in Wernher von Braun's FBI file:

Major James Hamill advised that in June, 1946 Magnus von Braun, the brother of Wernher, sold a bar of platinum to a jeweler in El Paso, Texas for $100.00. Magnus von Braun told the jeweler at the time of the sale that the platinum had been brought from Holland to the United States by his father, who had served in Europe during World War I. [The elder von Braun did not enter the United States until late March 1947, nine months after this incident]. The identity of von Braun was determined through his name and phone number which he had furnished to the jeweler. He was questioned by Major Hamill concerning this incident and readily admitted that he had brought the bar to the United States in violation of the Customs laws. Major Hamill stated that when the matter was brought to the attention of Wernher von Braun, as the director of the work by the German scientists, he had administered a severe physical beating to his brother.[40]

"German Scientists Plan Re-Fueling Station in Sky on Route to Moon" (*El Paso Herald-Post*, December 1946). Yes, the German rocket scientists had peaceful, although impractical goals.[34]

"We Want with the West" (*Time*, December 9, 1946). *Time* magazine published the first national report about von Braun and his team at Fort Bliss. The article was illustrated with a photograph of von Braun, standing before sections of his V-2 rocket, hands in pants pockets, appearing confident, proud, possibly arrogant, and triumphant, as if he worked for the side that had won the war—which, in fact, he now did. *Time* reported that the Germans were "civilian employees of the U.S. War Department, European Theater, on *temporary duty* [emphasis added] in the U.S." *Time* noted that they earned $2 to $11 per day and in addition received $6 per diem for detached duty. The article concluded, "Some day, they have been told, they may have a chance to become U.S. citizens. The fact that the U.S. is bringing their families to them [at Fort Bliss] seems to be a kind of guarantee that that is a promise."[41]

The El Paso Rotary Club invited Wernher von Braun, whom they now knew was the leader of the German rocket builders, to speak at their January 16, 1947, meeting.[42] Von Braun, who had been kept in obscurity for fifteen years, first by Nazi Germany and then by the United States Army, was delighted to go public. He prepared himself and his story.

Von Braun was acutely aware that he would have to surmount the language barrier. He practiced his English by speaking into a tape recorder, then played it back to hear how he performed. He worked to become fluent and to reduce his accent. He learned and used slang and colloquial expressions.[43]

Of course, von Braun could not tell the Rotary Club about his work for the Army; it was classified. He could, however, address the subject of "Future Developments of the Rocket." Von Braun described his agenda for venturing into space with well-developed descriptions of the equipment he wanted to build:

- an enlarged V-2 rocket as a "space ship"
- the design of a three-stage rocket that could lift an artificial satellite into orbit
- a winged rocket that could reenter the atmosphere and land safely
- a space station shaped like a wheel that produced an artificial gravity as it rotated
- the use of the space station as a base for trips to the moon and planets[42]

Von Braun might have hoped for a more influential forum, and his message was ahead of its time. Yet in his first public appearance in America, he defined the direction for his career. He would get a chance to tell his story again, although it would be five years before the American people would hear it.

The American southwest was a lonely place for educated Europeans who were a long way from home and struggling with a new culture. They wanted the company of their loved ones, and the Army obliged them. By March 1947, the wives and families of the German scientists arrived in El Paso.[44] Wernher von Braun was not to be left out as members of his rocket team settled into permanent residence and domesticity. The preceding fall, on November 7, 1946, he announced to the Army his engagement to Maria Louise von Quistorp with the request that his fiancée be allowed to accompany his parents from Landshut, Germany, to Fort Bliss, where the marriage would take place.[45]

Maria von Quistorp was the eighteen-year-old first cousin of Wernher von Braun. Her father, Dr. Alexander von Quistorp, was the brother of von Braun's mother, who had been christened Emmy von Quistorp.[46] Maria was an attractive blue-eyed blonde with the charm of the German aristocracy, of which she, like Wernher, had been a part. They last saw each other in the winter of 1945 when she was sixteen, and her family was hurriedly preparing to leave their home in northern Germany near the Baltic for a town near the Dutch border, to escape the advancing Soviet Army. Because of the circumstances, there was no time for a romantic discussion.[47]

In later years, von Braun told of how, at the age of seventeen, he held his infant cousin Maria in his arms at her baptism in the Lutheran Church. "That was the moment," he said, "when I looked into her eyes and decided to marry her."[48] It was a great story, but the kind best told to grandchildren. While von Braun reached the age of thirty-four without getting married, he had not escaped romantic involvements. At one time, he boasted that his "girlfriend" had been the legendary German aviatrix Hanna Reitsch.[49] They met as teenagers in the summer of 1932 when both took lessons in flying gliders at the Grunau Training School in Silesia.[50] While they remained friends throughout their lives, Hanna fell into a relationship with Luftwaffe Gen. Robert Ritter von Greim and became part of Hitler's inner circle during the final days of the war.[51,52]

Von Braun's most serious relationship seems to have been with Dorothee Brill, a twenty-seven-year-old Berlin woman, born in Tubingen, in the southwest of Germany. On April 5, 1943, von Braun submitted a formal application to the SS-*Rasse-und Siedlungshauptamt* (SS-RuSHA, the SS Central Office for Race and Settlement) for permission to marry Brill.[53] That is the first and only reference to Dorothee Brill in von Braun's records.

Had the SS-RuSHA found Jews among Brill's ancestors, thereby making her genetically unfit to marry von Braun? Did they get married? Was she a casualty of the constant air raids on Berlin? Or had love simply gone sour? The answer is hidden, as is much about Wernher von Braun's life before arriving in America, by his own choice, by the discretion of his friends and colleagues, through destruction of records as a result of the war, and as a result of the division of Germany that followed it.

In the months that followed the end of the war, Wernher von Braun's parents managed to find their way from the Soviet occupied zone to "Camp Overcast" in Landshut, in the American occupied zone of Germany. Maria von Quistorp was in the British zone of occupation, but was in contact with the von Brauns in Landshut.[54] Her father, Alexander von Quistorp, was unaccounted for as late as September 1948,[55] and was later reported to be in a prison camp in the eastern zone of Germany.[56] The elder von Braun became Maria's protector. His father brokered the marriage to ensure Maria's security and his son's happiness.[54, 57]

Like Wernher von Braun, many of the Germans brought to the United States were single men who also wanted the comforts of family and home life. Germany, destroyed, defeated, and divided among the victors, was full of single young women with no apparent future who were willing to come to the United States and marry men they hardly knew as a way of escaping bleak futures. The Army, however, was not inclined to go into the mail-order bride business for their German DASEs, including von Braun. If von Braun and any of the others wanted to get married, they could do so in Germany, and bring their wives back to Texas with them.

Wernher von Braun left for Germany on February 14, 1947.[58] He and Maria were married in the Lutheran Church in Landshut on March 1, 1947.[59] The marriage made news in El Paso two days later, when the *El Paso Herald-Post* printed the "**Story of Wernher von Braun's marriage, encompassing his experience in rocket development in Germany and his family background.**"[44]

Wernher von Braun returned to Fort Bliss on March 26, 1947, accompanied by his young wife and his parents. Since his brother Magnus was already there, most of the von Brauns had become immigrants. Only brother Sigismund, a former diplomatic officer to the Vatican, remained in Germany, where he was employed as a translator by the International Military Tribunal in Nuremberg.[60]

Along with the happiness Maria brought into his life, marriage to her was a good career move. One of von Braun's German colleagues, Ernst Stuhlinger, said the following about her: "From the very beginning, Maria was greatly admired by everybody for her youthful beauty and grace, and was deeply respected because she played her role as '*First Lady*' with so much dignity."[48]

The story of the German rocket experts, as recorded in the El Paso press, continued to show an amazing progression:

"**German Pupils Sing 'Eyes of Texas'; Like to Recite Pledge to U.S. Flag**" (*El Paso Herald-Post*, August 5, 1947).[61]

" 'Speak English' Contests Help Americanize Children of Germans" (*El Paso Herald-Post*, August 6, 1947).[62]

"Scientists Not Seeking Citizenship" (*El Paso Times*, July 27, 1947). Responding to criticisms that the German rocket scientists were applying for U.S. citizenship, a State Department spokesman stated that none had applied for citizenship and that none would be eligible to do so for another two years.[63] The story would change in just three months.

"German V-2 Scientists Here Among 72 Asking Citizenship" (*El Paso Times*, November 5, 1947). Seventy-two Germans in Project Paperclip applied for citizenship. Major Hamill said that every German at Fort Bliss who was eligible to apply did so, and that all of those who were not yet eligible, nonetheless, intended to renounce their German citizenship.[64]

In less than one year, the press—especially the El Paso newspapers—transformed von Braun and his rocket team from a bunch of enemy aliens into respectable immigrants, family men, and candidates for U.S. citizenship. They became real people, regular guys, at least as much as certified eccentric geniuses could be.

Of course, not all Americans were willing to forgive and forget that the Germans who came to the United States under the sheltering umbrella of Project Paperclip had been the darlings of the Nazis. Vehement protests began soon after the German scientists first met the press. On December 30, 1946, a group of eminent individuals that included Albert Einstein, politician Richard Neuberger, union organizer Philip Murray, and religious leaders Rabbi Stephen Wise and Norman Vincent Peale sent a statement of protest to President Harry S. Truman that included the following: "We hold these individuals to be potentially dangerous carriers of racial and religious hatred. Their former eminence as Nazi party members and supporters raises the issue of their fitness to become American citizens and hold key positions in American industrial, scientific, and educational institutions."[65]

On March 24, 1947, the executive secretary of the Federation of American Scientists, W. A. Higenbotham, also addressed President Truman with a demand that the Project Paperclip scientists be denied jobs in private industry or higher education. He wrote: "Any favor extended to such individuals, even for military reasons represents an affront to the people of all countries who so recently fought beside us, to the refugees whose lives were shattered by Nazism, to our unfortunate scientific colleagues of former occupied lands, and to all those others who suffered under the yoke these men helped to forge."[66] The Federation of American Scientists believed that the "wholesale importation of scientists as not in keeping with the best objectives of America and foreign policy."[67]

The July 1, 1947, *El Paso Times* reported, "**German Scientists in El Paso Blasted.**" Congressman John D. Dingell, Democrat from Detroit, denounced on the floor of the House of Representatives the Paperclip Program and those it brought to the United States.

Dingell said, "I have never thought that we were so poor mentally in this country that we have to go and import those Nazi killers to help us prepare for the defense of our country.

A German is a Nazi and a Nazi is a German. The terms are synonymous."

While Dingell's words may seem harsh over a half century later, he was repeating what was common belief and national policy little more than two years earlier.

The *El Paso Times* article went on to give the following disturbing comment:

> The British [press] warned that some of the scientists had perfected rockets which had killed British women and children and that some had committed war crimes more serious than those for which other Nazis had had to hang.
>
> For months the New York Times in articles has been warning Americans that among the scientists at El Paso were some very active Nazi party members.[68]

The article cited no specific instances of war crimes. It gave no names of active Nazi Party members.

For balance, the article noted the Army's position: The rocket scientists had been Army employees for over a year and showed themselves "technically and morally to the satisfaction of the War Department." Consequently, "wider avenues of research have been opened to them."[68]

The United States Army got the message. If it wanted to keep its German scientists in the United States, it had better show them to be model immigrants, or at least separate them from their Nazi pasts.

2

The Authorized Biography

I was still a youngster in my early 20's and frankly didn't realize the significance of the changes in political leadership. My father was wiser. He had been what corresponds to Secretary of Agriculture under President von Hindenburg, but quit all public offices when Hitler came into power. He warned me that it was all going to end in tragedy for Germany and many other people too. But I was too wrapped up in rockets to heed his warning.

Wernher von Braun[1]

Like many celebrities, Wernher von Braun carefully controlled what the public knew about him and his past. His story has been told many times. He gave full cooperation to many of his biographers and reviewed many of their manuscripts for accuracy. On a few occasions, he even told his story himself.[1,2,3,4,5] When it suited his needs, he bent the facts, occasionally told convenient lies, but why not? Politicians, movie stars, and immigrants did it all the time.

When he came to America, most of what was learned about Wernher von Braun, his ancestors, and his activities for Nazi Germany was told by von Braun himself, his friends and associates, and public records—what still existed of them. Much remained unknown because records were destroyed as a result of the war and, possibly, through premeditation. If any records still existed, they were not available to the United States Army because they originated and remained in the Soviet zone of occupation. What is most intriguing about von Braun's story before he embarked on his new life in America, is not the story that he told or created out of his imagination, but what he left out.

The American immigrant has been, for the most part, a commoner, a peasant, a black sold into slavery. A few were of the aristocracy, fallen on hard times, in search of their fortune in a new land. Wernher von Braun was of the latter group.

The von Braun ancestral name has been traced back to Ritter (knight) Henimanus De Bruno, who lived in the Bavarian city of Branau in 1285. Over the centuries, the family name has been spelled as Bruno, Brunowe, Bronav, de Bronne, Brawnaw, and Braun. De Bruno's descendants passed through the centuries as landowners with estates in Silesia and East Prussia. Wernher von Braun's father, Magnus Alexander Maximilian Freiherr (Baron) von Braun (1878–1972), continued the family tradition by owning estates in both East Prussia and Silesia.

Wernher von Braun's mother was born Emmy von Quistorp (1886–1959). While her ancestors cannot be traced as far back as the von Brauns, they were, nonetheless, prominent in Germany. The von Quistorp family originated in Sweden, but for several hundred years was present in Pomerania and Mecklenburg as ministers of the Lutheran Church, and as university professors, bankers, and landowners.

Magnus Freiherr von Braun married Emmy von Quistorp in 1910. The following year, Emmy gave birth to their first son Sigismund. A year after, on March 23, 1912, she gave birth to Wernher Magnus Maximilian Freiherr von Braun in Wirsitz in Posen province. In 1919, the von Braun's produced a third son, Magnus.[6]

Wernher was born two years before the start of the First World War. At the time, Baron von Braun busied himself with his land holdings and his position as *Landrat*, or provincial councillor for Posen. The war ended disastrously for Germany and for the von Brauns. In the final settlement, the German province of Posen was ceded to Poland as were, not incidentally, Baron von Braun's properties in the province.[3] The family then took as its primary residence the Baron's estate in Loewenberg county in Silesia. Far from the capital of Berlin, the von Braun's lived a life sheltered from the political and economic turmoil that churned through its cities during the 1920s. In the decade following the end of World War I, Baron von Braun acquired political influence when he became minister of agriculture for the Weimar Republic. He packed up his family and moved to Berlin.[6]

Wernher von Braun dated his interest in science and engineering to the day of his confirmation in the Lutheran Church. As a gift to commemorate the occasion, his mother gave him a telescope. The rest, as he described it, was inevitable: "So, I became an amateur astronomer which led to my interest in the universe which led to my curiosity about the vehicle which will one day carry a man to the moon."[3] The vehicle was, of course, the rocket.

Wernher von Braun was introduced to the rocket by two publicity seeking Germans, Max Valier and Fritz von Opel. Valier wrote about space travel and the rocket as its driving force. Opel was an automobile builder with a sense of the dramatic and an appreciation for a good publicity stunt. Valier enlisted Opel as his partner in financing his experiments with rockets. In the mid-1920s, solid-fueled or "powder" rockets were used for lifting signal flares or firing lines for rescues at sea. Valier and Opel bought off-the-shelf powder rockets and began bolting them to racing cars, gliders, and ice boats. When they set off the rockets, they set speed records and got publicity for Opel's autos and Valier's vehicle for space travel.[7]

When young Wernher learned of Valier and Opel's stunts, he went to a neighborhood fireworks dealer in Berlin and purchased a half-dozen sky-rockets. He tied them to his coaster wagon, and took his vehicle to the Tiergarten Allee, a major thoroughfare. He lit the fuses, the wagon took off out of control down the street pushed by a flaming tail, pedestrians scattered in panic and the police arrived to take the novice rocket scientist into custody. Fortunately, there were no casualties, and young von Braun was released to the custody of the minister of agriculture.

The wealth and prestige of his father and their residence in the capital opened doors for young Wernher to the best education Germany had to offer. His parents sent him to the French Gymnasium (high school) in Berlin, where all courses were taught in French. He learned French quickly, claiming he inherited his mother's gift for languages. However, he failed both mathematics and physics. Baron von Braun was clearly displeased; he sent his underachieving son to the Hermann Leitz Boarding School near Weimar, which had a reputation for fostering new teaching methods, close relationships between students and teachers, and a demanding curriculum.

One day while at boarding school, Wernher came across an advertisement in an astronomy magazine for a book he remembered as *Road to the Planets* (this was probably *Wege zur Raumschiffahrt*, or *Road to Space Travel*, published in 1929) by a man named Hermann Oberth. The images of a gigantic rocket and a distant moon dominated the ad. Wernher ordered the book, knowing that it would tell him how to travel into interplanetary space. When the book arrived and he looked at its pages, he was stunned. The book was filled with mathematical equations and tables of data. He would never be able to understand how to get to space—unless, of course, he learned mathematics and physics. Now, having an incentive and seeing the applications of math and physics, Wernher dove into his studies. He eventually surfaced with passing grades and graduated.[3]

In the Spring of 1930, Wernher von Braun returned to Berlin to enroll as an engineering student in the *Technische Hochschule* (Technical University) of Berlin, Charlottenburg.[8] In the last years of the 1920s, Germany in general and Berlin in particular was a place where young men dreamed and planned of building rockets to take them into space. They had organized the *Verein fur Raumschiffahrt* (Society for Space Travel, or VfR) as a means of pursuing this dream. When he arrived in Berlin, Wernher became active in the VfR. Through the group, he met a young writer named Willy Ley. In later years, Ley became the primary historian of German rocketry. He gave a concise description of Wernher von Braun, the college student: "Physically he happened to be a perfect example of the type labeled 'Aryan Nordic' by the Nazis in the years to come. He had bright blue eyes and light blond hair and one of my female relatives compared him to the famous photograph of Lord Alfred Douglas of Oscar Wilde fame. His manners were as perfect as rigid upbringing could make them."[9]

Willy Ley knew everybody in Germany who had more than a passing interest in rockets, and he introduced Wernher to the patriarch of German rocketry, Hermann Oberth, whose book fired von Braun's imagination in his last year of high school. Oberth was in Berlin for the purpose of testing a rocket motor he had designed. The introduction, it seems, was done over the telephone, and von Braun seized the opportunity.

"I'm still in technical school," von Braun said to Oberth, "and can't offer you anything more than spare time and enthusiasm, but isn't there something I can do to help?"

Oberth, who was financing his experiments by donations and out of his own pocket, also seized the opportunity. "Come right over," he said, and so, Hermann Oberth became Wernher von Braun's first teacher in rocketry.[3]

Hermann Oberth was born in 1894 in Transylvania, a remote corner of the Austro-Hungarian Empire. Following in his father's footsteps, he chose to study medicine, and he enrolled at the University of Munich. World War I interrupted Oberth's education, and when his service with a field ambulance unit was over, he gave up his interest in medicine. Also, with the end of the war, Transylvania was ceded to Rumania, an enemy of Germany; and Oberth became an alien, if not to the German culture, then to Germany. Nonetheless, Oberth returned to Germany to continue his studies in mathematics and physics. For his doctoral thesis, Oberth conducted an independent, theoretical study of rockets as vehicles for space travel, a subject that had fascinated him since childhood. The faculty at the University of Heidelberg rejected the thesis, probably due as much to their deficiencies of imagination as to actual faults with Oberth's theories.

Oberth would be bitter forever about not receiving a fair hearing by the German scientific establishment, but he refused to let his ideas of space travel die. He paid a publisher to print his thesis under the title *Die Rakete zu den Planetenraumen* (*The Rocket into the Interplanetary Space*). Amazingly, the slim volume was read widely, and Oberth found himself with a group of young disciples who wanted to build his rockets, the members of the VfR. The attention he received from outside the scientific mainstream sustained Oberth; and in 1929 he published a much expanded version of his book under the title *Wege zur Raumschiffahrt* (*Road to Space Travel*), which had caught young Wernher von Braun's attention.[10]

Also, in 1929, Oberth took a leave of absence from his job as a high school teacher in Rumania to travel to Berlin, where he entered into a strange alliance with Germany's great movie director Fritz Lang. Lang was creating a movie titled *Frau im Mond* (*The Girl in the Moon*), which was about a manned trip to the moon by rocket. To give his film authenticity, Lang hired Oberth and Willy Ley as technical advisors, and he convinced Oberth to build and fire a rocket on October 15, 1929, the day of the movie's premiere. *Frau im Mond* was a grand success; however Oberth's rocket was never completed. While he was a brilliant theoretician, he lacked the practical skills needed to build it.[11]

In 1930, Oberth returned to Berlin to try again to build and test a liquid-fueled rocket motor. He gathered around him as assistants a few members of the VfR, including Wernher von Braun. This time he gave up the grand plans that had baffled him when he worked with Lang, and designed a simple rocket motor he named *Kegelduese*, or "cone jet." He and his assistants succeeded in firing the motor in a controlled test sponsored by the Reich Institute for Chemistry and Technology (a body with responsibilities similar to those of the United States Bureau of Standards). When the test was over, Oberth had a certificate defining the performance specifications of his engine, the first liquid-fueled rocket motor to be developed in Germany. Although he held success in his hands, Oberth was once again without funds. He returned to his teaching job and obscurity in Rumania. Oberth left further development of the rocket in Germany in the hands of his disciples in the VfR and to Wernher von Braun.[2,12]

The young German rocket enthusiasts continued to pursue their interests through the VfR under the leadership of a World War I combat pilot and engineer named Rudolf Nebel. Nebel, who had an entrepreneurial spirit, arranged for the VfR to lease from the city of Berlin a former ammunition dump in the city's northern suburbs that would be the home, rocket building, and testing facility for the VfR. In late September 1930, they took possession of the property and hung a sign at its entrance that read "RAKETENFLUGPLATZ BERLIN," Rocket Airdrome Berlin.[2,13]

Wernher von Braun, Rudolf Nebel, and Willy Ley were among the dozens of rocket enthusiasts who assembled experimental rockets from scraps and fired them at the *Raketenflugplatz*. Their creations were imaginative, but unsophisticated and dangerous. Occasionally, they flew as intended. Their activities attracted the attention of the residents of Berlin, the local fire department, the press, and the German Army.[2,14]

The German Army, the *Reichswehr*, developed an interest in rockets. The Treaty of Versailles, which limited the future size and armaments of the German military, made no mention of rockets, which saw no significant use in World War I. To the Army, the rocket would be a permissible weapon with superior performance to that of conventional artillery—if it were developed. In the spring of 1932, several Army officers—in civilian attire, of course—visited the *Raketenflugplatz* to examine its developments. They were impressed by the work that had been done without significant financial support, but they were disappointed by the VfR's cavalier approach to documentation of design and performance. To find out if the *Raketenflugplatz* group could really develop a rocket, the Army offered to pay them 1,360 marks to build one and fire it from an Army artillery range. Implicit in the Army's proposal was the promise to support further work at the *Raketenflugplatz* if they produced a successful test rocket.

Early one morning in August 1932, Rudolf Nebel, Wernher von Braun, and a third rocket enthusiast named Klaus Riedel brought their hopes and their rocket to the Army's artillery firing range at Kummersdorf, south of Berlin. They were greeted by Capt. Walter Dornberger, who had been given the responsibility by the Army for rocket development. The rocket was of the "one-stick repulsor" design: The rocket motor held the forward position near the device's nose, and trailed behind it narrow fuel tanks resembling a stick. When the rocket was launched, it rose smoothly for about 100 feet, then it pitched over and flew horizontally till it crashed in a nearby forest. The demonstration was a great disappointment for the *Raketenflugplatz* group and for the Army. They had no justification for supporting amateur rocket developers.[2,15]

Wernher von Braun, the college student, the young aristocrat, would not accept failure. He gathered up what little data he could find about the *Raketenflugplatz* rockets and approached Col. Karl Becker, chief of ballistics and ammunition for the *Reichswehr*. Becker welcomed von Braun warmly, and listened to his proposal. Then he offered von Braun and the *Raketenflugplatz* group a deal. The Army would support them if they would work in absolute secrecy. Rudolf Nebel, who was the most influential member of the group, objected to the condition of secrecy, and he wanted no part of the Army's offer.[2]

Becker then made von Braun a second offer. When von Braun received his bachelor's degree in the fall, the Army would underwrite his graduate study at the University of Berlin (where Becker also held the position of professor), if von Braun would study liquid-fueled rocket motors.[16] Wernher von Braun earned Becker's offer by virtue of his intelligence, his enthusiasm, and his persuasive ability; it also helped that the young man was well connected. His father, Baron von Braun, minister of agriculture for the Weimar Republic, was Becker's friend.[17]

Wernher von Braun would conduct the experimental part of his doctoral research at the Army's *Versuchstelle Kummersdorf* West (Experimental Station Kummersdorf West). When he reported for work on October 1, 1932, he had his bachelor's degree in hand and was just twenty-years old. He was soon joined by a mechanic named Heinrich Greunow and another young rocket enthusiast named Walter Riedel (not related to Klaus Riedel). This team, all civilians, would work under the direction of Walter Dornberger, who had recently been promoted to Colonel. From the very start, Dornberger was the operational manager of the group, and within a very short time, Wernher von Braun, the technical genius, became the group's technical director.[18]

Years later, Wernher von Braun explained why he and his contemporaries chose a path that would lead to ever deepening involvement with the Nazis:

> We needed money for our experiments, and since the [German] army was ready to give us help, we did not worry overmuch about the consequences in the distant future. Besides, in 1932 the idea of another war was absurd. The Nazis were not then in power. There was no reason for moral scruples over the use to which our researches might be put in the future. We were interested in only one thing— the exploration of space. Our main concern was how to get the most out of the Golden Calf.[19]

Walter Dornberger became the second, and probably the most influential of Wernher von Braun's teachers and mentors. Dornberger was a career army officer. He served in the German Army during World War I as an artillery officer, and had the misfortune of being taken a prisoner of war just before the armistice in 1918. He spent the following two years in a French prisoner of war camp. After his release, he stayed with the Army while earning bachelor's and master's degrees. When he returned to active duty, he was assigned to the Army's ballistics branch with the assignment of developing the potential of rockets. Col. Walter Dornberger was thirty-seven-years-old, medium height, clean shaven, with blue eyes and thinning brown hair. He was self-assured, opinionated, and assertive; his character and temperament suited him for a career in the German Army.[20]

While von Braun, Dornberger, and their small group were devising the first, primitive rocket motors for the German Army, the political landscape of Germany was undergoing catastrophic change. Germany was in political disorder and economic depression since the end of World War I. Many, if not the majority of Germans, thought their country needed strong, nationalistic leadership. They got it. On January 30, 1933, Adolf Hitler became Chancellor of Germany.[21] The following March, the people of Germany handed over control of the Reichstag to the Nazis.[22]

In the dozen years that followed, according to his autobiographical articles and the accounts written by his friends and fans, Wernher von Braun devoted his time to developing rockets with the ultimate goal of building spaceships. Building weapons were a tangential accommodation he made for the privilege of pursuing his work. He was—according to these accounts—naive, disinterested in politics, and, except for the final year or two of the war, isolated from the intolerance and brutality that defined the Third Reich. According to Wernher von Braun, space travel was the passion that defined his life.

Of course, an interest in space travel and service to the Nazi cause are not necessarily mutually exclusive. In the Germany of the early 1930s, the former may have required the latter. If any nation or power could put men into space, it would be one with a clear national purpose, with a vision of its place in history, and with the will to make it happen. The authoritarian Third Reich of Hitler's Nazi Germany was the right patron to make space exploration possible.

The missiles von Braun's and Dornberger's group designed and built for the German Army were designated Aggregate; they were the sum of all German knowledge about rockets and their systems. By mid-1933, they had begun designing and building the Aggregate-1, the A-1. The A-1 looked like an artillery shell with tail fins. It was one foot in diameter and 4.6 feet long. It had a liquid-fueled motor that generated over 650 pounds of thrust, and it was to be stabilized in flight by the brute force of an eighty-five-pound gyroscopic flywheel in its nose. The A-1 was ground-tested and ready for flight in late 1933. A fraction of a second after the rocket motor was fired, the A-1 disintegrated into a ball of flame and shards of metal. The missile was the victim of delayed ignition of its engine.

Rather than build a second A-1, Von Braun and Dornberger set to work on an improved version, the A-2. It would have the same overall dimensions and rocket motor of its predecessor, but its gyroscopic flywheel would be located near the middle of its body, between the fuel and liquid oxygen tanks.[2,23]

While he worked on the A-2, von Braun completed his doctoral dissertation and submitted it to the faculty of the University of Berlin. The work, "Konstructiv, theoretische und experimentalle Beitraege zu dem Problem der Fluessigkeitsrakete" ("Constructive, Theoretical and Experimental Contributions to the Problem of the Liquid-Fueled Rocket"), was accepted on July 27, 1934.[24] It was immediately stamped secret by the Army and not published until after the war.[25] Nevertheless, at the age of 22, Wernher von Braun had the academic credentials and position to lead rocket development, not just in Germany but in the world.

In December 1934, von Braun made good on his promise when his and Walter Dornberger's team successfully launched two A-2 missiles, christened Max and Moritz, from the island of Borkum in the North Sea. Both missiles reached an altitude of about one and one-half miles.[2,23]

In 1935, American rocket pioneer Robert Goddard, working largely by himself with limited grant support, was also building and flying liquid-fueled rockets. He and von Braun were unaware of each other's work, and he had to independently solve the same technical problems encountered by von Braun. In solving them, Goddard developed lighter, longer, and thinner vehicles. He fired his A-series rocket on May 31. It reached an altitude of 7,500 feet above Roswell, New Mexico, thereby matching the performance of von Braun's A-2.[26]

Meanwhile, Wernher von Braun was leading the German Army's team to the next advance in rocket design, the A-3. While still a prototype, the A-3 would be a major technological leap over the A-2 and Goddard's A-series rocket. By the standards of the time, the A-3 was a monster. It was 2.3 feet in diameter, and 21.3 feet long. When loaded with fuel, it weighed 1,650 pounds and was driven into the sky by a 3,300-pound thrust motor. While its size allowed it to include all of the latest systems developments, and facilitated carrying instrumentation of all kinds, what set the A-3 apart from all of its predecessors was its guidance system. Unlike its predecessors, the A-3 did not hurtle toward its target on dead reckoning and a prayer; it had its own internal system that steered it, at least through its upward, powered flight.[2,27]

As von Braun sketched out the design of the A-3, the bureaucracies of the Third Reich took notice of the successful flights of Dornberger's and von Braun's A-2s. The German Army, which until then had given the Dornberger-von Braun rocket effort meager financial support, was now willing to commit millions of marks to their enterprise. The Luftwaffe also wanted to contract with the Army group to develop rocket power-plants for radical new aircraft. Ultimately, the Army pledged 6 million marks and the Luftwaffe 5 million marks more for the development of rockets and for building a new development and test facility at the remote site of Peenemuende on the Baltic Sea.[2]

With the funds committed, there was planning to do. The rocket base at Peenemuende would be a joint venture of the Army and the Luftwaffe, with the latter contributing its resources for the nuts and bolts of planning and construction. Walter Dornberger took the lead in defining the project that became synonymous with Peenemuende and with German rocketry. The Army, Dornberger reasoned, was no longer funding a research project with vague applications for the distant future. It wanted a weapon to give Nazi Germany a tactical edge over its enemies. Dornberger defined the specifications of the new rocket weapon. It should be capable of delivering a ton of explosive over a range of 160 miles. The weapon should fall within a half mile of its target—an accuracy twenty times better than conventional artillery. Its dimensions should allow it to be transported over roads or, if by rail, through any railroad tunnel.

Wernher von Braun, with Walter Riedel, one of the most experienced members of the group, had been sketching ideas for a large rocket. Dornberger's guidance now gave them focus and direction. The rocket weapon they envisioned in its first iteration would be over forty-five feet long with a diameter of more than five feet. It would have tail fins with a spread of almost eleven and one-half feet. It would require twelve tons of liquid oxygen and fuel and it would be lifted from earth by a rocket motor generating twenty-five tons of thrust. The rocket would reach 3,350 miles per hour and have a range of 172 miles. The German Army group designated its rocket weapon the A-4. The A-3, which was well along in the planning process, would be a platform for testing the individual systems and components that would go into the A-4. The draft specifications for the A-4 also guided the design of the workshops, test stands, and other facilities at Peenemuende.[28]

Construction of the base at Peenemuende proceeded faster than the building of rockets. Peenemuende was an isolated peninsula at the northern tip of Usedom Island, the western most of two large islands at the mouth of the Oder River, where it empties into the Baltic Sea. Because of its remoteness, Peenemuende was an ideal location for a secret rocket base, and because it was heavily forested, it was an ideal place to hide workshops and test stands. The Army occupied the western strip of the Peenemuende peninsula, and the Luftwaffe used the northwestern tip where it built an airstrip. Both sections were administered by the Army under the name *Heersversuchstelle Peenemuende* (Army Experimental Station Peenemuende, or HVP). The buildings that housed the staff and served as offices and design centers were simple and functional; they were one and two story structures with peaked roofs and few decorative embellishments. By May 1937, the Luftwaffe's crews finished the first phase of construction, and Army and Luftwaffe staff began to move in.[29,30]

Wernher von Braun became the technical director of HVP and served in the same position for organizations that succeeded it until Peenemuende was shattered by war and stripped of everything of value.[31]

Peenemuende was the ultimate playground for the *wunderkinder* who wanted to experiment with fireworks. They were doing what the amateurs at the old *Raketenflugplatz* had only dreamed about. They had their own remote enclave where they could build and fire rockets. True, von Braun and his team worked long hours, but it was a labor of love. For them it was, in reality, play.

Peenemuende also had the great outdoors as a recreational facility. Wernher von Braun's grandfather had gone hunting there, and so did Wernher von Braun and Walter Dornberger. When the weather was warm, the Baltic offered a pleasant site for sailing. And when the day was done, von Braun and Dornberger and the senior members of their team could relax and swap stories at the officers' mess.

On December 4, 1937, almost three years after the successful flights of Max and Moritz, the A-2 twins, the German Army team led by Wernher von Braun was ready to launch its new creation, the 21.3 foot long A-3. They chose as a launch site a tiny island about five miles north of Peenemuende called the Greifswalder Oie. The first test was a disaster. The rocket rose from the launch pad, made a quarter turn, pointed into the wind, deployed the parachute that was to return it to earth, then tumbled out of control until it fell into the sea. Wernher von Braun and his mentor Walter Dornberger puzzled over the failure for several days, then decided that the problems started with the premature release of the parachute. They removed the parachute and tried again. The second A-3 performed essentially as the first. A third A-3 was launched without a parachute and into a windless sky. It reached an altitude of approximately 2,500 to 3,000 feet before it, too, fell out of control and into the sea.

Clearly, some part of the design was defective. After extensive study, von Braun and his team pointed the finger of blame at the guidance system, which had been designed by the German Navy's top expert on gyro control mechanisms. They then began work on a replacement for the A-3, designated the A-5, which would have specifications similar to the failed A-3, but have an effective guidance system and be able to fly.[2,32]

By early 1939, the Luftwaffe caught on that being partners in the Army's rocket program was a very expensive venture. It decided to go its separate way. The Luftwaffe kept its airfield at Peenemuende and left the rocket development facility as the property and problem of the German Army. At that time, the Army facility went by the uninformative name of *Heeres-Anstalt Peenemuende* (Army Establishment Peenemuende, or HAP).[33]

March 23, 1939, should have been a very big day for Wernher von Braun. It was his twenty-seventh birthday, and it was also the day he would first meet Nazi Germany's Fuehrer, Adolf Hitler. Hitler had been persuaded to be personally briefed on the German Army's rocket development program. The briefing, however, would not take place at the elaborate development base at Peenemuende, but at the old Kummersdorf West facility, just seventeen miles south of the Fuehrer's Reich Chancellery at the center of Berlin. Walter Dornberger, as the Army officer in charge of developing liquid-fueled rockets, led a tour of the facility. He showed Hitler and his entourage the static firing of a 650-pound-thrust rocket motor and a 2,200-pound-thrust engine. They then met Wernher von Braun, who gave the Fuehrer a lecture on the systems and operation of their rockets using a cutaway A-3 as a model. This was followed by the static demonstration of an A-5, which had its skin and fins removed to show its systems in operation. Dornberger concluded the briefing by telling Hitler about the A-4, the advanced weapon that had yet to be built.

After discussing the German Army's Aggregate series of rockets over lunch, Hitler commented, "Es war doch gewaltig!" which translated as, "Well, it was grand!" or, in colloquial English, "That was swell."

Walter Dornberger took Hitler's comment as faint praise for the Army's rockets, but he may have been optimistic. Hitler may have been commenting on the quality of his lunch of mixed vegetables and mineral water.

Dornberger later expressed his surprise that Hitler did not seem impressed by the ear-shattering demonstrations of rocket motors, the complex rocket mechanisms, or the plans for the future. Hitler's skepticism was, of course, justified. Dornberger and von Braun had spent tens of millions of marks, yet they had not successfully fired a rocket for over four years, not since the launching of the A-2s in December 1934.[34]

The world changed irrevocably on September 1, 1939, when Adolf Hitler ordered German military forces, which had been poised for attack, to invade Poland. Within weeks, Germany and the Soviet Union, its temporary ally, divided the country. England and France declared war on Germany, and they and other nations began to fall into the abyss of war.[35]

Dornberger, von Braun, and the team continued their work, and in October 1939, a year after the failed launchings of the A-3, the A-5 was ready to be fired. It had similar dimensions to the A-3: 21.3 feet long, and 2.6 feet in diameter. It had the same 3,300-pound thrust liquid-fueled motor, but with added improvements and instrumentation, it weighed nearly 2,000 pounds.

Three A-5 prototypes were launched from the Greifswalder Oie. All three flights were stunningly successful. The rocket-power lifted the missiles on guided flight paths for forty-five seconds, then parachutes lowered them to soft landings in the sea, from which they were recovered.

With three strikingly impressive successes, nothing could stop the German Army, Wernher von Braun, and his rocketeers. They had the design for the weapon, the A-4 guided missile, and the Third Reich had the war. The funds for building the A-4 seemed assured.[2,36]

Hitler thought otherwise. The war was going well and only needed more conventional armaments to reach a successful conclusion for Germany. In February 1940, he ended all weapons development projects that would not reach production stage within a year.[37,38] The German Army kept its massive rocket program going at Peenemuende by diverting funds from other projects and classifying 4,000 skilled workers as serving on front line duty.[39]

Two and a half more years passed before the weapon Wernher von Braun and his team built, the A-4, was ready to be tested. The first A-4 missile was fired from Test Stand VII at Peenemuende on June 13, 1942. It rose drunkenly several thousand feet into a cloudy sky, then fell back to earth. The second A-4, fired on August 16, rose majestically for four seconds until its guidance system froze. It raced through the sound barrier on dead reckoning, then forty-five seconds into its flight at an altitude of 35,000 feet, it exploded.[40,41] The pressure for success of the third attempt on October 3 was as tangible as the thunder of the A-4's motor when it fired.

The A-4, nearly forty-seven feet tall and weighing fourteen tons, stood in the great dirt-walled amphitheater that was Test Stand VII, at the very northern tip of Peenemuende. Walter Dornberger and his military staff, and Wernher von Braun and his senior engineering staff watched from nearly a mile to the south as the countdown approached zero. They saw the rocket rise smoothly over the pine forest, and only after the sound waves traveled the distance from the test stand did they hear the rolling thunder as the power of the rocket motor throttled up to roar with full power. The missile rose vertically for 4.5 seconds, then slowly tilted to the east. After about twenty-two seconds, it punched through the sound barrier and kept accelerating. It continued its climb at an angle of fifty degrees from the vertical, which would give it maximum range, and it was leaving behind a white trail of condensed exhaust against the clear blue sky. After fifty-eight seconds of powered flight, ground control shut off fuel flow to the engine. The rocket was traveling at almost 3,500 miles per hour on a ballistic course to its target area in the Baltic Sea, 120 miles east of Peenemuende. Five minutes after liftoff, the missile crashed into the sea, marking its site of impact with the bright green dye that had been its payload.

Wernher von Braun and Walter Dornberger rushed from their observation points to Test Stand VII, where a spontaneous celebration was taking place. Hermann Oberth, von Braun's first teacher was there, too. He had worked his way back into Germany through a series of minor appointments to reach the Mecca of rocketry, only to find himself left far behind in the rocket exhaust of his former protégé. Still, Oberth received his share of congratulations and credit for having been the inspiration for it all.

Later that evening at a formal celebration, Walter Dornberger addressed his staff: "We have invaded space with our rocket, and for the first time ... have used space as a bridge between two points on earth; we have proved rocket propulsion practicable for space travel. To land, to sea, and [to] air may now be added infinite empty space as an area of future intercontinental traffic." Then he added, "So long as the war lasts, our most urgent task can only be the rapid perfecting of the rocket as a weapon."[42]

When the Third Reich handed out awards for the successful flight of the A-4, Wernher von Braun accepted the *Kriegsverdienstkreuz I Klasse mit Schwerten* (War Merit Cross, First Class, with Swords).[43]

Long after the defeat of Nazi Germany, von Braun wrote, "The rest of the A-4 story was less spectacular and certainly less pleasant, not only for those who were to sit at the receiving end of its operational use against London or Antwerp, but also for its originators."[44]

The war became unpleasant for all of Germany about a month and a half after the first success of the A-4. By November 1942, the German Sixth Army had pushed deep into the southeast of the Soviet Union, as far east as Stalingrad. On November 19, 1942, the Soviet Army began a counter-attack that by the end of January 1943 turned the war against Germany in the east. Of the 330,000 men who had been part of the German Sixth Army, only 100,000 survived to surrender, and only about 5,000 survived the Siberian prisoner of war camps to return home after the war.[45] Brothers, fathers, and husbands were lost in the silence of defeat, and the German people began to understand that the Third Reich would also be lost, first to Allied bombers, then to invading armies.

The triumph of the October 3, 1942, flight of the A-4, unfortunately, was not followed by more successes. The rocket was unreliable in its flights and often broke up while returning to earth. Von Braun and his staff worked to correct the missile's deficiencies. Walter Dornberger regularly toured the centers of power in Berlin with the hope of raising more funds to finish developing the A-4. He slowly attracted attention. In May 1943, Minister of Armaments Albert Speer and his advisors watched a successful firing of an A-4 from Peenemuende. Two days later, Speer informed Dornberger that he was promoted to Major General.[46] Reichsfuehrer of the SS Heinrich Himmler also watched test firings of A-4's from Peenemuende, and he offered to take the Army's case for top priority status to Hitler.

With the war going badly for Germany, Hitler began to show signs of desperation. He developed enthusiasm for those advanced projects he had ordered canceled early in 1940 when he thought he had won his war. On July 7, 1943, Maj. Gen. Dornberger received orders to report to Hitler's headquarters at the Wolf's Lair in Rastenberg, East Prussia, to brief the Fuehrer on the A-4. He brought with him Wernher von Braun and Ernst Steinhoff, head of the Department of Instruments, Guidance, and Measurements.

The three men from Peenemuende met the Fuehrer in a lecture hall at his headquarters. Hitler was accompanied by Field Marshal Wilhelm Keitel, chief of staff of the armed forces, Gen. Walter Buhle, chief of armaments for the Army, Albert Speer, and their personal aides. They took their seats and Wernher von Braun took the stage. The lights dimmed, and, with commentary by von Braun, a motion picture projector began to show a film of the first successful firing of an A-4 nine months earlier. The film then showed details of the rocket's support system: the assembly building at Test Stand VII, a rocket rolled into place for firing, a static test, the mobile launching system with the rocket carried on a *Meillerwagen*, the positioning of the missile in the field, and loading the A-4 with fuel. Then, in case

Hitler and his staff had not been impressed by the film's opening sequence, it repeated the first successful firing of the rocket into the blue sky over the Baltic.[47]

Albert Speer described von Braun's performance and its effect on the Fuehrer: "Without a trace of timidity and with boyish sounding enthusiasm, von Braun explained his theory. There could be no question about it: From that moment on, Hitler had been finally won over."[48]

When von Braun finished his technical presentation, Walter Dornberger explained the hard facts of field deployment and production. The only significant discussion was whether the A-4 would be fired by mobile field units or from massive, fixed bunkers. Dornberger favored the former, but Hitler favored the latter. The Fuehrer got his way, of course. The bunkers would be built. And Walter Dornberger got what he wanted, a top-priority ranking for the A-4 from the Fuehrer.[49]

After the meeting, Wernher von Braun was also rewarded for his service to the Third Reich. On a prompt from Walter Dornberger, Albert Speer approached Hitler with a proposal to confer upon von Braun a titular professorship. A titular professorship was not an academic position, but a civilian honor granted by the head of state. Hitler, caught up in the brilliance of his new wonder weapon, agreed. He would sign the papers, and Speer would then make the formal presentation to von Braun.[48,50]

As the German Army's rocket team finally gained the patronage of Adolf Hitler, it also attracted the attention of the Allies, especially British Intelligence. They put together a disturbing amount of information that Germany was developing strange new weapons at a remote research base on its Baltic coast. It sent an RAF Mosquito on a photo-reconnaissance mission over Peenemuende. When developed, the photographs revealed the elaborate, secret base hidden in the forest, and several weapons that Great Britain would soon face, including the A-4 missile.[51]

On the night of August 18–19, 1943, the RAF amassed 497 Stirling, Halifax, and Lancaster bombers in the moonlit sky over Peenemuende.[52] Their mission, which had the approval of Winston Churchill, was not simply to destroy the base, but to kill the scientists and engineers who worked and lived there, and whose creations threatened England.[53] The raid lasted forty-five minutes, and when it was over, a firestorm raged through Peenemuende. The primary targets of the bombing, the scientists and engineers, escaped largely unharmed. Of the approximately 4,000 Germans who lived on the base (Peenemuende's staff and their families), 178 were killed. The remainder of those killed, about 557, were foreign workers, unskilled Russians and Poles, who had been conscripted to do general labor. They were trapped in their camp, just south of the Peenemuende facility, when the bombs fell.

Damage to the facilities was scattered and far less than desired by the British. The Pre-Production Works—the pilot plant intended to manufacture all prototypes and a significant number of A-4s when they went into production—suffered only minor damage. The Works, however, continued to be vulnerable. Hitler quickly ordered that production of his wonder weapon be moved to a secret, underground factory in the Harz Mountains of central Germany, a facility known as Mittelwerk. Hitler gave the responsibility for excavating the tunnels and building the factory to Heinrich Himmler and the SS.[52,54]

For some time, Heinrich Himmler, Reichsfuehrer of the SS and head of the Gestapo, had been pressuring the Army and Walter Dornberger to get control of the rocket program. Wernher von Braun told a story—not confirmed by other sources—of how Himmler pressured him toward this end. In February 1944, Himmler telephoned von Braun to invite him to his headquarters at Hochwald in East Prussia. Von Braun recalled that he entered Himmler's office with considerable trepidation because he viewed the man as being "as mild mannered a villain as ever cut a throat," which history has shown to be an accurate evaluation of the Reichsfuehrer. However, according to von Braun, Himmler was "quite polite" and as intimidating as a country school teacher.

"I hope you realize that your A-4 rocket has ceased to be a toy," Himmler said, "and that the whole German people eagerly await the mystery weapon. . . . As for you, I can imagine that you've been immensely handicapped by Army red tape. Why not join my staff? Surely you know that no one has such ready access to the Fuehrer, and I promise you vastly more effective support than those hidebound generals."

"Herr Reichsfuehrer," von Braun replied, "I couldn't ask for a better chief than General Dornberger. Such delays as we're still experiencing are due to technical troubles and not to red tape. You know, the A-4 is like a little flower. In order to flourish, it needs sunshine, a well proportioned quantity of fertilizer, and a gentle gardener. What I fear you're planning is a big jet of liquid manure! You know, that might kill our little flower."[2]

On reading von Braun's account of his meeting with Himmler, one can only wonder how he could have been so flippant with the coldest cutthroat of the twentieth century. Long after von Braun told this tale, new facts emerged that give new perspective to his story and give rise to questions about his candor in telling it (see Chapter 3). Inexplicably, von Braun did not tell his friend and boss Walter Dornberger about his meeting with Himmler.[55]

Three weeks later, von Braun was arrested in the middle of the night by agents of the Gestapo. He and several other members of his staff, including his younger brother Magnus, were accused of treason. The Gestapo claimed that von Braun and his team put their dreams of space flight above developing the A-4 as a weapon for the Third Reich. They languished in an SS prison in Stettin for two weeks until the persistent lobbying of Walter Dornberger[56] and the intercession of Albert Speer won their release.[57]

While von Braun had been clearly caught up in a power play between the German Army and the SS, he emerged with the reputation of a man whose priority was space exploration, not service to the Nazi cause. His brief experience as an accused enemy of the Third Reich would serve him well after the end of World War II.

There is one puzzling aspect of the arrest of von Braun and his colleagues that neither he nor others who have written about it have addressed. The Gestapo was legendary for the consistent and systematic brutality with which it treated those it arrested. The Gestapo routinely roughed up and tortured its prisoners, not to obtain confessions, but to use every opportunity to extract information about subversive activities.[58] There are no reports that von Braun and his group were mistreated while in custody or that other arrests of Army or Peenemuende personnel followed, which might indicate they had informed on others. It appears that von Braun and his colleagues were pawns in Himmler's power play, and were protected so that they could be used again.

The fundamental relationships that Wernher von Braun had with the German Army, with his mentor Walter Dornberger, and with the thousands of men on his rocket team changed as a result of the actions of a disabled, and disillusioned German Army officer. Lt. Col. Klaus Philip Schenk Count von Stauffenberg had lost part of his leg, his right hand, part of his left hand, and his left eye to a land mine in Tunisia. When he was discharged from the hospital, he was assigned as chief of staff to Gen. Friedrich Fromm, commander in chief of the Home Army and head of Army armaments. In his position, Stauffenberg made regular status reports to Hitler at his Wolf's Lair headquarters.[59] On July 20, 1944, Stauffenberg entered the conference room where Hitler was presiding over a briefing. A few minutes later, he excused himself and departed, leaving his briefcase behind. The bomb inside Stauffenberg's briefcase killed one man outright, a dozen others were severely injured, and three would later succumb to their injuries. Hitler, the target of the assassination attempt, suffered burns, lacerations, and bruises. His eardrums were shattered and his right arm was temporarily paralyzed, but he was alive.

Stauffenberg's coconspirators in Berlin, all Army officers, were to seize power, but in the moment of crisis, they lacked the courage to act. By the end of the day, the SS rounded up the conspirators, including Stauffenberg, and summarily executed them.[60]

Stauffenberg's commanding officer, General Fromm, asserted his ignorance of the plot and his innocence, but he was arrested, nonetheless. Hitler transferred Fromm's authority and responsibilities to Heinrich Himmler. The Reichsfuehrer of the SS took control of the Home Army and the Army's weapons development branch, which included Walter Dornberger's and Wernher von Braun's rocket development program.[61]

Before Himmler could consolidate his new position, the Army moved to deprive him of the prize. It had been planning to spin off *Heeres-Anstalt Peenemuende* (Army Establishment Peenemuende, or HAP), as a civilian business enterprise owned by the Third Reich. On August 1, 1944, HAP became the *Elektromechanische Werke* (Electromechanical Industries, or EKW). Maj. Gen. Walter Dornberger lost his command and gave up control of the facilities and organization he spent twelve years building. The industrialist brought in to manage the new company had the good sense to see himself as a caretaker and leave the real work of management to the people who knew what the organization was doing. Thus, the highest ranking employee of the EKW, Wernher von Braun, became the de facto leader of German rocket development.[62]

While Himmler failed to gain control of the rocket development enterprise at Peenemuende, he still controlled the Mittelwerk, the factory that produced the A-4. In addition, in early September 1944, Himmler, through his subordinate SS Lt. Gen. Hans Kammler, succeeded in bullying the German Army to surrender command of field operations of the A-4.[63] Kammler was uniquely unqualified for the task, having had no front line military experience. He was an architect by trade and made his mark in the Third Reich by designing and building Auschwitz-Birkenau, including its gas chambers and crematoria,[64] by orchestrating the leveling of the Warsaw Ghetto after the uprising had been crushed,[65] and by building the Mittelwerk factory where the A-4 was built.[66] Nevertheless, on September 8, 1944, Kammler choreographed the first successful firing of the missile on London and the barrage of weapons that would subsequently fall on England and parts of western Europe.

The day after London came under fire, a German newspaper headline announced, *"Vergeltungswaffe-2 Gegen London im Einsatz"* (Vengeance weapon-2 in action against London.) Joseph Goebbels's propaganda machine transformed the A-4 into the V-2, the designation by which von Braun's rocket would be known to history.[67]

Years later, when von Braun became a resident of the United States, though not yet a citizen, he told of his reaction to the operational use of the V-2: "New arrivals at Peenemuende could not understand us. Fresh from one military defeat after another, they said, 'Be happy with your V-2. It's our only weapon that the Allies can't stop. It's a success. It's hitting London every day.'"

" 'It's a success,' we said, not too loudly, 'but we're hitting the wrong planet.' "[3]

Wernher von Braun continued to lead his group at Peenemuende in designing, and in one case, testing successors to the A-4 (V-2). The A-9, which for the sake of exploiting the A-4's top-priority rating was designated the A-4b, had the body of an A-4 to which were fused sharply swept-back wings. The purpose of the wings was to allow the missile to take a gentle glide path to earth rather than coast on a ballistic arc to its target. It could travel twice the range of the A-4 (V-2) with a flight time of seventeen minutes. On January 24, 1945, an A-9 (A-4b) launched from Peenemuende reached a speed of 2,700 miles per hour. Although it broke up on its return glide to earth, it was the first, albeit unpiloted, supersonic airplane.

The next step in development, though it never went beyond the drawing board, was the A-10. The A-10 was envisioned as a reusable booster that would push the A-9 to 2,700 miles per hour. The winged A-9 would then travel under its own power, reaching a speed of 6,300 miles per hour, then glide back to earth. In a forty-minute flight, the two-stage A-9/A-10 vehicle could carry a thousand pound payload a distance of 2,500 miles, the distance from northern Europe to New York.[2,68]

In von Braun's mind there were bigger and bigger boosters, possibly designated A-11 and A-12, that might lift the A-9 and even a thirty-ton A-10-sized vehicle into orbit around the earth.[2] During the winter of 1944–45, as Germany was being destroyed, these designs were dreams; the A-4 rocket and the team that designed and built it were also in danger of being destroyed.

By the end of January 1945, the workers at Peenemuende could hear the booming cannons of the Soviet Army fifty miles away. There was no doubt in anybody's mind that the rocket base would inevitably fall to the enemy. Wernher von Braun called a clandestine meeting of a few of his senior staff to discuss their options. He took a straw vote and received a unanimous response: von Braun and his German rocket team would not wait at Peenemuende to be captured by the Soviets, but would cross Germany to deliver themselves and their expertise to the Americans. The choice of the Americans was simple. The United States was the only Allied power that had, in their estimation, the resources and the will to continue development of their rockets. The decision to surrender to the Americans was kept secret within the small group since it was an act of treason against the Third Reich.[3,69]

On the last day of January, von Braun summoned section chiefs, department directors, and members of his inner management circle to his office. He announced that he had just received an order from SS Lt. Gen. Hans Kammler to evacuate personnel and equipment involved in their most important projects to central Germany. Von Braun emphasized that it was an order, not a proposal.[70] Von Braun later admitted that he had on his desk several conflicting orders from various segments of the Third Reich, and he took the one that best suited his goals.[3,69]

Not surprisingly, von Braun's staff quickly organized the exodus from Peenemuende of about 3,000 staff, equipment of any value, and tons of documents relating to rocket design, construction, and testing. They moved by railroad, trucks, and even barges along Germany's rivers. By early March 1945 most of the move was completed. Von Braun set up his offices at Bleicherode.[71] Walter Dornberger, who had been shifted aside into performing support and advisory activities, set up his headquarters at Bad Sachsa, also in central Germany.[72] Both towns were within a short distance of the subterranean Mittelwerk factory, where the V-2 missiles were being built.

There was no longer any chance that von Braun and his team could continue rocket development, or for that matter, that Nazi Germany would survive. His goal now was to keep his team together. One night in mid-March, von Braun's car raced up the autobahn to Berlin for a meeting at the Ministry of Armaments where he planned to plead for funds to build a new laboratory. The possibility of raising the money or building the laboratory was unlikely, but the charade of asking for it kept alive the notion that von Braun still led a cutting-edge technological team. Von Braun never reached Berlin. His driver fell asleep at the wheel. Von Braun crawled from the wreckage with a shattered left shoulder and his left arm broken in two places. He spent the next few months with his arm and shoulder in a cast, physically exhausted, futilely trying to keep his team together from his makeshift headquarters in Bleicherode.[73]

Telling the story years later, von Braun made the following puzzling comment about his situation: "we there fell into the hands of a local tyrant who was the most ruthless man I ever met. He was one of Himmler's SS generals named Kammler."[1] The comment was disingenuous because von Braun worked with Kammler—though not necessarily willingly—for a year and a half; and he knew that Kammler was anything but a local warlord. Besides his influence in the operation of the Mittelwerk factory, Kammler controlled the field operations of the V-1 buzz bomb[74] and von Braun's A-4, the V-2 ballistic missile.[75] Von Braun was well aware of Kammler's character, if not the specific details of his service to the Reich, before he chose to follow Kammler's order to relocate the Peenemuende establishment to central Germany.

By early April 1945, American tanks had pressed to within a dozen miles of von Braun's headquarters at Bleicherode, and they were about to overrun the entire area surrounding the Mittelwerk factory.[76] Kammler ordered that von Braun and a core group of 400 of his top scientists again relocate, this time to Oberammergau in the foothills of the Bavarian Alps.[77] Walter Dornberger and his small staff received the same order.[78] Kammler's motives in issuing these orders are ambiguous. Superficially, it appeared that Kammler was ordering the group into the Alpine Redoubt, an area of safety in Germany's impenetrable southern mountains where the Third Reich could regroup and continue the battle. Secretly, Kammler hoped to negotiate with the Americans to surrender Germany's aircraft and rocket technology for a price, a plan remarkably similar to von Braun's.[79] It is not known if von Braun knew of or agreed with Kammler's plans, but he was in no position to refuse the order to relocate.

On or about April 11, General Kammler—he had been promoted at the end of March—summoned Wernher von Braun to a meeting where he told von Braun that he would be leaving Oberammergau to deal with other responsibilities of his to the Reich, and that von Braun and his team would be left in the care of his subordinates. The following day, von Braun verified that Kammler had left Oberammergau; and, with the exception of a flurry of messages to SS headquarters, Kammler disappeared from history forever.[80]

In the days that followed, von Braun's rocket team dispersed through the villages surrounding Oberammergau, where they pursued make-work projects, awaited the end of the Third Reich, and minimized the chance of all them being killed at once by an Allied air raid or an SS purge. Von Braun sought attention for his injured shoulder and arm.[81]

On May 1, 1945, German radio announced—embellishing the truth—that the Fuehrer, Adolf Hitler, died during combat at his headquarters in Berlin. The following day, May 2, Wernher von Braun and six other members of the German rocket team, including his younger brother Magnus and his mentor Walter Dornberger, crossed the Alps into Austria and surrendered to the United States Army.[82]

In the days after their surrender, the German rocketeers were discreet in what they told their captors. The group of seven and other members of the team who the Americans rounded up were held at Garmisch, where they divulged only what von Braun, General Dornberger, and Dornberger's chief of staff, Lt. Col. Herbert Axster approved. They did not want the United States Army to take everything they had and send them back to the ruins of the Fatherland. They wanted to trade their assets for the best possible deal they could get,[83] and this reluctance to tell all was noted with some resentment by von Braun's interrogators.[84]

Under the direction of Col. Holger Toftoy, the United States Army began to strip the Mittelwerk of enough parts to build 100 V-2 rockets. The Americans also located the fourteen tons of documents von Braun ordered hidden for safekeeping, and they began rounding up the cream of von Braun's staff. With the rocket parts, the documentation, and the rocket scientists, the United States Army had the beginnings of its own rocket program.[85]

Wernher von Braun remembered these pivotal days in his life: "Officers at the U.S. Army detention center in Germany questioned me severely for several weeks. Finally a colonel [Holger Toftoy] put this blunt question to me: 'Do you think you could become a loyal citizen of the United States?'

"I said I would like to try."[1]

Years later, Dr. Richard Porter, who played a major role in the interrogations of von Braun and the German rocket builders, was asked who originated the idea of bringing von Braun and his team to the United States. He answered, "Probably von Braun himself as much as anyone."[86]

Wernher von Braun's story was well known, but surprisingly few bothered to ask the obvious questions:

- Why did he so quickly switch his allegiance at the end of World War II?
- Von Braun admitted that he had joined the Nazi party to protect his career. Did he simply use the Nazis, and vice versa, or was he one of them?
- What did von Braun know about the concentration camps and what went on inside them?
- How did von Braun and his team of rocket engineers, who worked for the Nazis, enter the United States so quickly after the end of World War II, while refugees and survivors of concentration camps waited for years to get the same opportunity?

Or, to squeeze these questions into one all-encompassing puzzle: What did Wernher von Braun omit in his autobiographical articles and in his authorized biographies?

3

The Cover-Up

> Screen them for being Nazis? What the hell for? Look, if they were
> Hitler's brothers, it's besides the point. Their knowledge is valuable for
> military and possibly national reasons.
>
> United States Army officer[1]

Operation Overcast threatened to start a war within the United States government. Overcast was a War Department activity. The War Department's original intent of short-term exploitation of the Germans' technical knowledge was expanding into long-term contracts and the importation of greater numbers of scientists and technicians than needed to fulfill its original objective. The War Department plundered Germany of its technical experts for the benefit of the United States military and related industries with the express purpose of keeping the experts out of the hands of the Soviet Union. Overcast brought hundreds of enemy aliens and their dependents into the United States without the benefit of passports or visas.[2] Not surprisingly, the State Department found this ad hoc approach unsatisfactory.

In early 1946, the State-War-Navy Coordinating Committee formed and drafted a policy that would permit legal entry of the Germans and allow them to be hired on long-term contracts. On March 16, 1946, Operation Overcast opened up into the broad umbrella of Project Paperclip.[3] The policy statement governing Project Paperclip was signed by President Harry S. Truman on September 6, 1946.[4]

Even with Project Paperclip sanctioned as official policy, the immigration of the Germans to the United States was a tricky proposition. Various federal laws forbade the immigration of those who were members of Fascist groups or opposed United States efforts during the war. Paperclip policy made their immigration more problematic because it stated that "No person found . . . to have been a member of the Nazi Party and more than a nominal participant in its activities, or an active supporter of Nazism or militarism shall be brought to the U.S. hereunder."[4] The policy, of course, gave the military some wiggle room in assessing whether an individual could qualify for immigration. It further stated that "neither position nor honors awarded a specialist under the Nazi Regime solely on account of his scientific or technical abilities" would prevent him from participating in Project Paperclip.[4] Translation: A candidate, such as Wernher von Braun, could have racked up accolades from Hitler and the perks of rank in the Nazi establishment without being automatically denied admittance to the United States.

The process of legitimizing the immigration of German specialists and scientists had the Joint Intelligence Objectives Agency (JIOA) the agency under the Joint Chiefs of Staff that was responsible for Project Paperclip, submit dossiers on their candidates to the Departments of State and Justice for their approval. The key document in each dossier was a report from the Office of the Military Government U.S. (OMGUS), which administered the defeated Germany. The OMGUS report was to contain any information available in Germany relating to membership in Nazi organizations and responsibility for war crimes. In the first few rounds between JIOA and the State Department, the latter took a hard-stance line against candidates with Nazi backgrounds, categorizing them as security risks in view of their ideological position during the recently ended Second World War. The paper trail left behind by the process of approving Wernher von Braun's entry into the United States revealed how the Army covered up the inconvenient facts of his Nazi past.[4]

On April 23, 1947, OMGUS had in hand its first "records check" of von Braun's membership in Nazi organizations. This document indicated the following:

- Military service in the Luftwaffe: May 1, 1936–June 15, 1938. (Von Braun took pilot training on two occasions, and apparently spent most of this time detached to the German Army which employed him at Peenemuende, where the Luftwaffe also had a base.)
- Joined the Nazi Party: May 1, 1937.
- Was a member of the SS with ranks conferred on the following dates:

Untersturmfuehrer (SS second lieutenant)	May 1, 1940
Obersturmfuehrer (SS lieutenant)	November 9, 1941
Hauptsturmfuehrer (SS captain)	November 9, 1942
Sturmbannfuehrer (SS major)	June 28, 1943.[5]

Von Braun's appointment as a *Sturmbannfuehrer* was a double honor for him, since evidence indicates that Heinrich Himmler personally announced the promotion and pinned the symbols of his new rank on von Braun's SS uniform.[6,7] Not only was von Braun's SS membership damning in view of the role of the SS in running concentration camps and the extermination of the Jews, but von Braun's rank of Sturmbannfuehrer had significant stature. Adolf Eichmann's rank in the SS was only one grade higher.

What did von Braun have to say about his Nazi record? On June 18, 1947, he wrote an affidavit explaining his memberships and a bit more.

While a student at the Institute of Technology in Berlin, I joined the DVL ("Deutscher Luftsport-Verband," German Sport Aviation Club) in order to become admitted to the aviation school at Berlin-Staaken. There I made my pilot's license in summer 1933. . . . In summer 1934, the club was absorbed by the N. S. Fliegerkorps (National Socialist Aviation Corps). All members of the DVL were taken over automatically by the latter. I remained a member of the Fliegerkorps until approximately summer 1935. [Von Braun continued flying when he entered military service with the Luftwaffe the following year.]

In fall 1933, I joined the SS horseback riding school on the "Reitersturm I" at Berlin-Halensee. I was there twice a week and took riding lessons. The entire outfit did never participate in any activity whatever outside the riding school during my connection with it. In summer 1934, I got my discharge from the Reitersturm.

In 1939 [his record, however, indicates the year was 1937], I was officially demanded to join the National Socialist Party. At this time I was already Technical Director of the Army Rocket Center at Peenemuende. . . . My refusal to join the party would have meant that I would have to abandon the work of my life. Therefore, I decided to join. My membership in the party did not involve any political activity.

In spring 1940, one SS-Standartenfuehrer (SS Colonel) Mueller . . . looked me up in my office at Peenemuende and told me, that Reichsfuehrer SS Himmler had sent him with the order to urge me to join the SS. I called immediately on my military superior for many years in the Kriegsministerium (War Department), Major General Dr. Dornberger. He informed me that . . . if I wanted to continue our mutual work, I had no alternative but to join.

After having received two letters of exhortation from Mueller, I finally wrote him my consent. Two weeks later, I received a letter reading that Reichsfuehrer SS Himmler had approved my request for joining the SS and had appointed me Untersturmfuehrer (lieutenant).

From then on I received a written promotion every year. At the war's end I had the rank of "Sturmbannfuehrer" (Major). But nobody ever requested me to report to anyone or to do anything within the SS.

In addition to the SS I was a member of the following organizations: DAF (Deutsche Arbeitsfront, Trade Union); NSV (Nat. Soc. Welfare Organization); Deutsche Jagerschaft (Hunting Org); Reichsluffschutzbund (Air Raid Protection Organization).[8]

Von Braun's affidavit said, in short: I was a member of many organizations. I had to be to keep my job. I was not an active member of the Nazi party or the SS; I was a member in name only. In the years that followed, von Braun stuck to his story that, although his heart was not in it, he was compelled to join the Nazi party. He would not admit to an association with the SS that started in 1933, soon after Hitler came to power, and ended with him holding the rank of SS major.

The fact that Wernher von Braun was a major in the SS gives new perspective to his story of being offered a position by Heinrich Himmler, declining the position, and subsequently being arrested by the Gestapo. Himmler personally honored von Braun and, whether or not von Braun agreed with the details, Himmler lent his support to the mass production of the A-4/V-2 missile. Himmler apparently had reason to expect a more cooperative attitude from the young rocketeer.

OMGUS submitted its original report for inclusion in von Braun's dossier on September 18, 1947. The document signed by Col. C. F. Fritzsche read, in part:

> Based on available records, subject is not a war criminal. He was an SS officer but no information is available to indicate that he was an ardent Nazi. *Subject is regarded as a potential security threat* [emphasis added] by the Military Governor, Office of Military Government for the U.S. [OMGUS]. A complete background investigation could not be obtained because subject was evacuated from the Russian zone of Germany.[9]

Clearly, this incomplete report on von Braun would have invited the State Department to blackball him. Other candidates had similar problems, and the JIOA would have to do something about these damaging reports if they hoped to get the State Department to approve the immigration of the German Scientists. On December 4, 1947, the JIOA director, Bosquet Wev, asked the European Command director of intelligence for help:

1. OMGUS security reports recently forwarded from your headquarters classify (14) specialists [including Wernher von Braun] as potential or actual threats to the security of the United States. . . .

3. There is very little possibility that the State and Justice Departments will agree to immigrate any specialist who has been classified as a potential or actual security threat to the United States. This may result in the return to Germany of specialists whose skill and knowledge should be denied other nations in the interest of national security.

4. It is requested that the cases of the specialists listed in paragraph one be reviewed and that new security reports be submitted where such action is deemed appropriate in view of the information submitted in this letter.[4]

Thus, the JIOA determined that the German Nazis who had been judged security risks should be admitted to the United States in the interest of national security.

OMGUS was apparently a team player when it came to the desire of the military versus the judgment of the Departments of State and Justice. The revised security evaluation of Wernher von Braun, again signed by Col. C. F. Fritzsche and dated February 26, 1948, read:

Further investigation of Subject is not feasible due to the fact that his former place of residence is in the Russian Zone where U.S. investigations are not possible. No derogatory information is available on the subject individual except NSDAP records, which indicate that he was a member of the Party from 1 May 1937 and was also a Major in the SS, which appears to have been an honorary commission. The extent of his party participation cannot be determined in this Theater. Like the majority of members, he may have been a mere opportunist. Subject has been in the United States more than two years, and if, within this period, his conduct has been exemplary and he has committed no acts adverse to the interests of the United States, it is the opinion of the Military Governor, OMGUS, that *he* [von Braun] *may not constitute a security threat to the United States* [emphasis added].[10]

OMGUS and JIOA apparently did not relay to the Departments of State and Justice what they had in their files on von Braun's involvement in Nazi organizations.

That Wernher von Braun's security report was whitewashed to facilitate his entry into the United States was apparently not an isolated case. Linda Hunt, a reporter who studied Project Paperclip documents, examined 130 OMGUS reports of German specialists, many of whom were part of von Braun's rocket team. She found that all 130 reports were revised to remove suggestions that the subjects were security threats.[4]

The full dossier on Wernher von Braun was, to say the least, awkward. But did his memberships in Nazi organizations prove he was a Nazi? In Germany under Hitler, Germans were expected—some say required—to join the Nazi Party as a show of patriotism, to keep their jobs, and to keep out of concentration camps, which many also claimed they knew nothing about. After all, what is a Nazi? Was it a matter of membership cards, or was it a matter of acceptance of the Nazi ideology and actions in support of the Nazi cause?

The major Nazi war criminals were tried at Nuremberg between November 1945 and September 1946. With the convictions and punishments resulting from this trial, most of the world felt free to forget lesser war crimes and ignore trials resulting from them. Yet the number of accused war criminals ran into the thousands. The U.S. Military Court in Germany tried a total of 1,672 defendants in 489 cases from 1945 to 1949. The record of *United States of America v. Kurt Andrae et al.*, commonly known as the Dora-Nordhausen War Crimes Trials, tells of crimes committed in the production of Wernher von Braun's A-4/V-2 rocket. The trial record encompasses six trials held by the U.S. Military Court all relating to crimes allegedly committed at the Mittelwerk factory, the Dora concentration camp, and at its associated camps. Nineteen defendants were tried jointly, and five additional defendants had separate, quicker, and shorter proceedings. The court case was named alphabetically after the first defendant (Kurt Andrae). The court was in session at the former Dachau concentration camp from August 7, 1947 to December 30, 1947.[11]

The shameful story of the Mittelwerk and Dora war crimes began in September 1943, after the RAF bombed Peenemuende. The Nazi hierarchy resolved to build an impregnable underground factory to manufacture the Third Reich's wonder weapons, the V-1 flying bomb and the Army's A-4/V-2 rocket. The factory's location was in an expanded mine under the scenic Harz Mountains of central Germany and was owned by a company named Mittelwerk, G.m.b.H., or Central Works, Limited. In fact, the company was government owned and headquartered in Berlin, and financed through Albert Speer's Ministry of Armaments. The SS would also exercise substantial control over the factory and supply labor for construction and operation of the Mittelwerk.[12]

The digging of the tunnel system that housed the Mittelwerk, the building of the factory in it, and the actual construction of the V-weapons were all done by forced inmate labor, slave labor controlled and directed by the SS. The slaves worked, slept, and lived—if that word can be used—in the tunnels. For the inmates, the tunnels quickly became a gallery of the twin terrors of slow death resulting from insufficient nutrition and nonexistent sanitation and from sudden death at the murderous hands of the SS.[13]

Albert Speer, the man responsible for financing the Mittelwerk, toured his fledgling enterprise on December 10, 1943. After inspecting the tunnels for about an hour, Speer recalled the

> expressionless faces, dull eyes, in which not even hatred was discernible, exhausted bodies in dirty gray-blue trousers. At the approach of our group, they stood at attention upon hearing a cutting command and held their pale blue caps in their hands. They seemed incapable of any reaction. . . . The prisoners were undernourished and overtired; the air in the cave was cool, damp, and stale and stank of excrement. The lack of oxygen made me dizzy; I felt numb.[14]

Speer later claimed that to improve conditions he approved the allocation of materials for the construction of a respectable concentration camp above ground.[14] The subterranean concentration camp was moved and given the name "Dora." Dora existed only to supply slave labor to the Mittelwerk and associated factories to build the V-1 buzz bomb and the V-2 ballistic missile.

The Allies discovered Dora and the Mittelwerk when United States Army troops entered the area on April 11, 1945. They unearthed a human disaster that had been planned by the Nazi regime, executed by the SS, and compounded by the economic collapse of the Third Reich in the final weeks of the war. The victorious army discovered corpses everywhere. Hundreds of victims had been dumped on the ground adjacent to the crematorium, a low building part way up a hill. In the days before liberation, thousands died of starvation and cholera. Bodies were fed into the ovens four at a time, day and night, but piled up faster than the ovens could consume them. At the nearby city of Nordhausen, the Nazis also created an associated camp to house even more slaves for the Mittelwerk. It, too, was littered with corpses.[15,16] By one estimate, it held 6,000 bodies in varying stages of decomposition. During their two years of operation, 60,000 inmates passed through the Mittlewerk and the Dora, Nordhausen, and associated concentration camps; at least 25,000 died there.[17]

All twenty-four defendants in the Dora-Nordhausen War Crimes Trials were indicted under the general charge of "violation of the laws and usages of war." They were also charged specifically with acting "in a common design" to operate the concentration camp and subterranean factory that resulted in mass starvation, beating, torture, and murder of prisoners. The prosecutors also maintained that the defendants' disregard for the prisoners' needs for food, shelter, sanitation, and medical care constituted criminal behavior. All defendants pleaded not guilty to both the general and specific charges against them.[11]

The defendants in the trial were mostly low ranking SS officers and soldiers, and "Kapos," concentration camp inmates, mostly common criminals, who were recruited by the SS to help keep the other prisoners in line. The Kupos were assigned to do the dirty work in the camp and the factory, and the ones most likely to commit crimes in the presence of witnesses. The highest ranking member of the group was an SS captain, who was outranked by Wernher von Braun. The lone civilian manager on trial was Georg Rickhey, general director of the Mittelwerk factory.[11] Rickhey arrived in the United States on July 4, 1946, as a Project Paperclip specialist employed by the Army Air Force. Four months later, another Paperclip specialist accused Rickhey of responsibility for the mass hangings of prisoners at the Mittelwerk. Rickhey was returned to Germany to stand trial.[4]

The chief defense counsel for Georg Rickhey, Maj. Leon B. Poullada, demanded that Wernher von Braun and other Paperclip rocket specialists who worked with Rickhey and were familiar with the operation of the Mittlewerk, return to Dachau to testify in Rickhey's defense. The demand seemed reasonable since Poullada's client faced a possible death sentence if convicted. Army Ordnance staff flatly refused to let von Braun or any of its German DASEs return to Germany to testify. There was no way to control the testimony and prevent damage to the German technical teams being built in the United States under Project Paperclip. The best defense counsel Major Poullada could get were depositions from von Braun and several others, which he could enter into evidence.[18] His questions made it clear to von Braun and the others what answers were wanted. They focused on testimony favorable to Rickhey and probed no deeper. They did not delve into areas in which von Braun and the others might incriminate themselves.

Wernher von Braun gave his deposition on October 14, 1947. (The questions were numbered. Not all questions and answers are included here.):

1. *Q.* What is your full name?
   A. Wernher von Braun.
2. *Q.* What is your residence?
   A. Fort Bliss, Texas.
3. *Q.* What is your occupation?
   A. Project Director.

4. *Q*. What is your educational background?

A. Ph.D. in physics (Dr. phil (Physik)), University of Berlin, Research Professor

5. *Q*. Briefly what has been your professional experience and qualifications?

A. As Director of Development of the Heeresversuchsanstalt Peenemuende-Ost (later on Elektromechanische Werke G.m.b.H) I was responsible leader for the development of the A-4-rocket (V-2).

10. *Q*. Did you yourself work in the Mittelwerk factory at Nordhausen? If so, please give the dates.

A. No.

11. *Q*. If you were not employed in Mittelwerk did you ever visit the factory? If so, what dates and in what capacity?

A. Yes, I visited the later Mittelwerk for the first time in September or October 1943. . . . Later on, after A-4 production was in operation, I have been there 15 to 20 times, approximately, for discussing technical matters in connection with the technical alterations of the A-4. The last time I was there was in February 1945.

12. *Q*. During your visit did you observe general working conditions in the Mittelwerk factory from May 1944 to April 1945?

A. Working conditions at the Mittelwerk were continuously improved during the entire period from the last months of 1943 up to my last visit at the plant. In the beginning, such conditions were extremely primitive, since the tunnels . . . were not fit in any way for starting accurate production and to absorb many thousands of workers. Since a camp was not available, the prisoners were housed in the tunnels proper under the most primitive conditions. However, in the summer of 1944 considerable improvements had been made or were partly under construction.

18. *Q*. Is it correct that Rickhey was in effect just a figure head and the real authority was held by Sawatzki?

A. Yes. Sawatzki was given exclusive authority in all matters concerning management.

20. *Q*. Is it correct that overall production programs were set by the Production or Armaments Ministry and that administration of such programs was in the hands of Sawatzki?

A. The work programs of the Mittelwerk were determined by the leaders of the competent production commissions of the Speer Ministry. Sawatzki being a member of such commissions was personally responsible to their leaders for the feasibility of the programs assigned to the Mittelwerk.

21. *Q*. Is it correct that Sawatzki had great power and influence in

Mittelwerk because of his connections with SS General Dr. Kammler and because of his position as a member of the Special Board for V-weapons?

A. In the fall of 1943 SS General Kammler was given the command by Himmler to develop . . . a subterranean V-weapon production plant (V-1 and V-2) with the assistance of his construction organization made up of prisoners from concentration camps. After July 20, 1944 (Attempt to assassinate Hitler) Kammler was successful in getting assignment with the command of troops trained in V-weapons. . . . Kammler [also] succeeded in dictatorial manner to get hold of the production. He passed up Rickhey entirely and gave his commands directly to Sawatzki who, very expertly but with the utmost ruthlessness against the prisoners as well as the German engineers and skilled workers, tried to carry them out.

33. Q. Is it correct that many civilian workers worked side by side with the prisoner workers of Mittelwerk under exactly the same conditions of temperature, air, etc?

A. Yes. I would estimate that the number of civil engineers and civil skilled labor amounted to at least 1,000 men. Besides, a larger number of female civil administration aides working on average of 12 hours daily, was employed in the tunnels and working in wooden shacks set up for this purpose inside the tunnels.

40. Q. Is it correct that for all matters of normal discipline and control the prisoners were exclusively under the guard of the SS?

A. As far as I know, yes.

41. Q. Is it correct that because of top secret nature of Mittelwerk project, all matters dealing with sabotage, espionage and subversive political activity, the prisoners were disciplined and punished by the SD [Sicherheistdienst, Security Service of the SS] or the Gestapo?

A. If I remember correctly, it was done exclusively by the SD.[19]

For a man who professed to be not involved in the activities of the Mittelwerk, Wernher von Braun knew a great deal about its operations. His testimony supported his colleague Georg Rickhey's claim to innocence of war crimes at the Mittelwerk. Von Braun's involvement in and responsibility for war crimes at the Mittelwerk never came up, but his position as an officer in the SS, which ran the Dora concentration camp and was eventually responsible for production of the V-2 missile designed by von Braun and which controlled the field operations of the weapon, left him in a very awkward position. Von Braun needed to identify villains, which he did in the persons of Albin Sawatzki, production director of the V-2 missile, and SS Gen. Hans Kammler, who directed construction of the underground rocket factory, exercised significant control over V-2 production, and directed field operations of the missile. Sawatzki and especially Kammler were demonstrably culpable for war crimes at the Mittelwerk, but they were also ideal scapegoats: their whereabouts in 1947 were unknown.

Albin Sawatzki was in American custody in Nordhausen, Germany, on April 14, 1945, when he gave a deposition on his activities at the Mittelwerk. He claimed that he was technical director of the Mittelwerk factory and was promoted to factory director only four weeks earlier.[20] He pointed the finger of guilt for the atrocities committed there at the SS. While there were several reports of his whereabouts afterwards, none were confirmed.[21] Likewise, Hans Kammler disappeared amid numerous stories of his glorious death in defense of the Third Reich or in suicide after its defeat.[22]

Several other members of the Project Paperclip group at Fort Bliss also gave depositions in support of Georg Rickhey's defense. They were fed the same questions that were posed to Wernher von Braun, and to a large extent, they gave similar answers. Ironically, as the testimony at the trial unfolded, responsibility for war crimes shifted from Georg Rickhey to another Paperclip specialist. While defending Rickhey and covering-up for each other, one witness accused Arthur Rudolph, the chief operations director of V-2 production, of turning over the names of inmate-saboteurs to the SD (the security force of the SS). When Rudolph gave his deposition to Rickhey's defense counsel, he was a resident of Fort Bliss, Texas.[23]

The military court that tried *United States of America v. Kurt Andrae et al.* announced its verdict on December 30, 1947. Fourteen defendants were found guilty and given prison terms of up to twenty-five years (none of whom actually served more than eleven years); one defendant was sentenced to death and hanged; four defendants, including Georg Rickhey, were acquitted.[11] Although acquitted, Rickhey did not return to the United States to continue working under Project Paperclip.

After Project Paperclip completed its security clearance report on Wernher von Braun in 1947, the records were sealed and classified by the United States Army until November 1984,[24] six years after von Braun's death. After the completion of the Dora-Nordhausen War Crimes Trials, and after the sentences were carried out, the trial transcript was classified by the United States Army until it appeared in the National Archives in 1981.[11]

The worms were back in the Army's can, and the Army sealed it tight. The United States Army's official position was that Wernher von Braun had done nothing wrong during the Nazi regime in Germany. The embarrassing facts were hidden and the crimes committed in the production of the V-2 rocket and von Braun's complicity in them were officially forgotten; in fact, they were never common knowledge. When von Braun and his colleagues wrote their biographies and histories of the development of the V-2 rocket in Germany, they rarely mentioned the Mittelwerk and never mentioned the Dora concentration camp or the atrocities committed there. The only ones who remembered were the survivors of Dora and its associated concentration camps. And they remained silent for many years until Wernher von Braun and his German rocket team became national heroes of the country that liberated them and defeated the Nazis.

Fifty years after the fact, we can ask the question: Who was responsible for the cover-up? Who chose to ignore, or ordered others to suppress records of the Nazi activities of Werner von Braun and other German specialists brought to the United States under Project Paperclip? Who obstructed a full investigation of the crimes committed at Dora then buried the records of the war crimes trials? The documents that exist have the signatures of a small number of minor Army personnel and bureaucrats. Their job was not to bring Wernher von Braun and his rocket team to the United States, but to legitimize the status quo. Von Braun and his team were already at Fort Bliss and White Sands when the cover-up took place in 1947 and 1948.

When Wernher von Braun and six others surrendered to the United States Army at the end of the war, the young lieutenant who had them in custody was told to find out if they were Nazis. "Screen them for being Nazis!" he said in disbelief. "What the hell for? Look, if they were Hitler's brothers, it's besides the point. Their knowledge is valuable for both military and possibly national reasons."[25] This attitude permeated Army thinking and it became unofficial policy.

It has been said that history is written by the victors. In the case of the German rocket program, history was written by the defeated. In 1947, the United States Army knew only what von Braun and his fellow Germans told them in their depositions, and the Army was not interested in digging deeper. Though there was damaging information to be found, unearthing it would be left to historians decades later. Wernher von Braun wrote extensively about his life in Nazi Germany in articles and books, as author and coauthor. In only one work did he mention the Mittelwerk as the site of V-2 production.[25] He never wrote about Dora, the use of concentration camp slave labor, or the crimes committed in the production of his rockets.

The crimes of the Nazi era in Germany, according to von Braun, were the moral responsibility of Adolf Hitler, Heinrich Himmler, and SS zealots like Hans Kammler. The German Army was innocent, and, of course, von Braun and his colleagues were only interested in building rockets for space travel. He failed to mention that he had been a *Sturmbannfuehrer* in the SS, or admit any responsibility for the crimes committed in the production of his V-2 ballistic missile.

4

The Reborn Rocketeer

> Von Braun, like Moses, led his expatriates through the desert toward a distant promised land.
>
> Thomas O. Paine[1]

Wernher von Braun and his rocket team surrendered to the United States with ambitious plans their new masters did not share. The team was the world's leading expert on designing, building, and launching big rockets. They knew it, as did the United States Army, the Soviets, the British, and the French, but the United States government, with the exception of a clique within the Army, did not care. The war against Germany and Japan was over, and there was no need for radically new weapons systems or the costs associated with them. Von Braun and his team were housed in the desert to keep the capability available if it was ever wanted. They described their situation with phrases such as "prisoners of peace"[2] and as being "on ice in the desert."[3] Their euphoria of being at the cutting edge of rocketry was over.

Life at Fort Bliss was exile, frustration, boredom, and lack of respect. Von Braun and his team existed in "rat shacks": wood frame barracks surrounded by barren hills, sand, rattle snakes, and buzzards. El Paso was the last vestige of the old frontier minus the romance. Beyond it, across the river was Juarez, with all the evils of poverty. The launch site at White Sands, deeper in the desert, was worse than Fort Bliss. The Texas and New Mexico desert was not home for von Braun and his German team, and it would never be. It was at best a refuge where they would hold out until the Army needed a new weapon or became interested in space exploration.

The Army showed the value it placed on Wernher von Braun and his team by the man—or more accurately, the rank of the man—they placed in charge. At Peenemuende, they worked under the command of Maj. Gen. Walter Dornberger; at El Paso, they worked for Maj. James P. Hamill. Though nobody had the bad taste to mention it, von Braun's rank in the now defunct SS was the equivalent of Hamill's.

Hamill was a lanky, twenty-six-year-old native of New York City. He had been a student at Fordham University, where he was a physics major and member of the Reserve Officers Training Corps (ROTC). When he graduated in 1940, he entered the Army where he was assigned to Ordnance Technical Intelligence. In the final days of the war, his commanding officer, Col. Holger Toftoy, assigned him the task of removing 100 V-2 rockets from the Mittelwerk and shipping them to White Sands for examination.[4] At Fort Bliss, Major Hamill had a handful of junior officers and enlisted men to help him run the Office of the Chief of Ordnance, Research and Development Service, Suboffice (Rocket), and to keep over one hundred German engineers and scientists under control and productive. Major Hamill reported to Colonel Toftoy, head of Army rocket development, whose office was in Washington, D.C.[5]

White Sands was under the command of Col. Harold R. Turner. Operations there were conducted by his staff of officers and enlisted men, employees of the General Electric Company, with the occasional participation of Navy staff.[6]

When the novelty of America wore off, reality sunk in. Von Braun admitted frankly that he was disappointed by what he encountered at Fort Bliss. At Peenemuende, he and his team were pampered by the German Army and supported with the vast resources of the Third Reich. In America, which was demobilizing its armed forces and cutting military expenditures, they were reduced to advancing a rocket program on the Army's meager dole.

The jobs often seemed pointless, and the Army expected results without paying for them. Much of the work involved assembling and firing V-2s and training the Americans in their operation. The missiles, which were shipped from the Mittelwerk as parts, had deteriorated with age. Von Braun's team often had to scrounge for materials and fabricate new parts in order to put together functioning missiles. Arthur Rudolph complained bitterly about the shortage of metric wrenches. Von Braun also felt that

Major Hamill's authoritarian style of dealing with the German staff undercut his authority over his team.[7]

Von Braun saw himself and his team wasting away in the desert with no future. He had held together his rocket team, which he turned over to the Americans, and they in turn did not give him the facilities and support that his team needed to advance the science of rocketry. Von Braun regularly gave up in disgust. Major Hamill received von Braun's resignations personally, by telephone, and in writing. Hamill's position was that the Army had von Braun and every German member of what he considered *the Army's rocket team* under contract. None of them were free agents. Hamill ignored von Braun's verbal resignations and threw the written versions in the trash.[7,8]

Even as the Army compelled its German rocket group at Fort Bliss to subsist on marginal support, it took credit for their rocket developments and expertise. Fort Bliss, Texas, which was established to anchor the territorial interests of the United States, celebrated its one hundredth anniversary in November 1948. While it spent most of its existence as a sleepy frontier outpost, it celebrated its centennial with an eye toward a new frontier. To commemorate its founding, the United States Post Office issued a three-cent postage stamp that showed a rocket lifting off. The missile was remarkably similar in design and scale to the V-2.[9] Ironically, the V-2s were flown from White Sands, not Fort Bliss, and there was no evidence on the stamp of von Braun and the Germans who came to Fort Bliss with him.

While Wernher von Braun's career stagnated and his dreams of space travel languished, he faced new responsibilities as head of his extended family. In Germany, he was a bachelor, without responsibility for anyone other than himself, with the freedom to enjoy everything life in the Third Reich had to offer. In America, von Braun had a young wife to support, and also counted as dependents his elderly parents who lived with him. While his younger brother Magnus was formally independent and an adult, Wernher had brought him to the United States as part of his rocket team and saw to it that Magnus had a livelihood, if not a future with the Army. In addition, von Braun claimed as dependents Irmgard and Henrike Riedel, the widow and daughter of Klaus Riedel, his close friend and colleague from their days with the *Raketenflugplatz* group in the early 1930s. Riedel died in an automobile accident in July 1944, and von Braun's generosity allowed Riedel's wife and daughter to live comfortably at the camp for dependents run by the Army at Landshut in Germany.[10] Von Braun became irrevocably tied down by domesticity and personal responsibility on December 9, 1948, when his wife Maria delivered their first child, a daughter, Iris Careen.[11]

Wernher von Braun was born into a Lutheran family, and in Germany he viewed himself as a nominal Lutheran. Religion and morality were not high priorities in Germany for either the Third Reich or for von Braun. His life was full, with long hours and, in the end, frantic, health-destroying activity. America seemed the opposite, with little progress in rocket development but with more time for family. As the Third Reich receded in time and space and with the documentation of his life in Germany either destroyed or classified by his employer, the United States Army, Wernher von Braun's sins were wiped away, if not forgiven. He was in a position to recast his values, re-create himself, to be reborn as a believer in God, a proponent of democracy, and a visionary of mankind's potential in the heavens.

He wrote of his first years in America:

> In El Paso I found there were a great number of small churches housed in temporary barracks. And I saw pastors getting their congregations together in buses and trucks. To me it all indicated a terrific pioneer spirit.
>
> I frankly had never considered myself the church-going type before coming to America, and was skeptical of the need for any organized religion. But in the American towns where I went I found myself admiring the work churches were doing in making Christianity alive in the community, and became grateful for the job they did in helping our German families feel at home, including my own.[12]

Von Braun claimed he attended church services regularly.[13]

Of science and technology, he said: "All our harnessing of nature is evil if it serves only to enslave man to the machine or to the political organizations that control the machines. What the world most needs today, to my mind, are people who will get out and shout for human freedom and the rights of individuals."[13]

Von Braun was preaching to the choir: With the defeat of Nazi Germany, the enemy of freedom was Communism and the Soviet Union.

After marriage and before children, after the workday and before bedtime, after the Third Reich and before television, Wernher von Braun had time to fill and ambition to spare. His dreams of space travel were on hold until his new patron, the United States Army, put a value on space. Until that happened, he filled his time with a technical exercise, which became *The Mars Project*. In his youth, he had been inspired by Jules Verne's book *From the Earth to the Moon* (1865), and he began to write a novel about a similar voyage through space to the planet Mars. Like Verne, von Braun rigorously based his story on the known science and technology of the day, extrapolating it in scale and into the future. His message was that not only was space travel possible, it was inevitable.

In his introduction to *The Mars Project*, in its final, published form, von Braun wrote:

The study will deal with a flotilla of ten space vessels manned by not less than 70 men. Each ship of the flotilla will be assembled in a two-hour orbital path around the earth, to which three-stage ferry rockets will deliver all the necessary components such as propellants, structures and personnel. Once the vessels are assembled, fueled, and "in all respects ready for space," they will leave this "orbit of departure" and begin a voyage which will take them out of the earth's field of gravity and set them into an elliptical orbit around the sun.

At the maximum solar distance of this ellipse which is tangent to the Martian orbit, the ten vessels will be attracted by the gravitational field of Mars, and their rocket motors will decelerate them and swing them into a lunar orbit around Mars.[14]

Von Braun went on to describe the landing of an exploration party on the surface of the red planet, and the party's return to earth. The entire scenario was outrageously ambitious, minutely detailed, and fantastic. To give his imaginative novel credibility, von Braun wrote an eighty-page technical appendix that worked through the engineering requirements of the rocket-powered vehicles that would make the voyage possible.

Then, in 1948, when he completed his manuscript, he sent *The Mars Project* to a New York publisher. It was returned six weeks later with a polite letter of rejection. Seventeen publishers and seventeen rejection letters later, von Braun accepted the reality that for whatever reason, no one was interested in publishing his manuscript.[15]

The multiple rejections must have been doubly frustrating for von Braun when he saw the publication in 1949 of *The Conquest of Space*, the immensely popular book by Willy Ley, his former colleague in the VfR and the man who had introduced him to Hermann Oberth. Ley's book was lavishly illustrated with paintings by Chesley Bonestell, some of which depicted rocket ships bearing a striking resemblance to von Braun's A-4/V-2 design.[16] While the author and illustrator were apostles of the creed of space, neither was its prophet. Neither ever designed or built a vehicle capable of entering space as did von Braun and his team while serving Nazi Germany. Still, Ley's and Bonestell's book supported von Braun's agenda. Another amateur, Arthur C. Clarke, would write about it decades later, "*The Conquest of Space* probably did more than any other book of its time to convey to a whole generation the wonder, romance and sheer *beauty* of space travel."[17]

As von Braun was planning the great adventures beyond earth, he still tended to business for the Army. The V-2 firings became as much—or possibly more—an American enterprise as an activity of von Braun's group. The Army, working with its contractor, the General Electric Company, as part of the Hermes program, tested new designs of components on rockets fired from White Sands. These components found their way into several prototype Hermes rockets, though they never reached operational status. The Navy had for some time been interested in launching large rocket weapons from its vessels, and it test-fired a V-2 from the deck of the aircraft carrier *Midway* on September 6, 1947. Subsequently, the Navy conducted its own rocket development program separate from the Army.[18]

The most spectacular and possibly the most significant flights of the V-2 in America were as part of the Army's Project Bumper, which was conceived as a way of testing methods of firing two-stage missiles and of sending payloads to extremely high altitudes. Project Bumper's rocket vehicle was composed of a V-2 first stage and a Wac Corporal second stage, which carried an instrumentation package. At the time, the Wac Corporal was the most sophisticated rocket developed by the United States. It was created by a team at the California Institute of Technology's Jet Propulsion Laboratory for Army Ordnance. When fueled, it weighed a diminutive 655 pounds in comparison with about 28,000 pounds for the V-2.[19] On September 26, 1945, the same month von Braun and the first contingent of Germans arrived in the United States,[20] the Wac Corporal attained a top speed of about 2,800 miles per hour[19] and reached a respectable altitude of nearly forty nautical miles (seventy kilometers).[20]

Between May 13, 1948 and July 29, 1950, the Army conducted eight Bumper launchings, six at White Sands and the last two from what was then called the Long-Range Proving Ground at Cape Canaveral, Florida. According to Wernher von Braun, the only completely successful missile of Project Bumper lifted off from White Sands on February 24, 1949. The second stage reached a velocity of 5,150 miles per hour and an altitude of 244 miles.[18] Project Bumper had squeezed all the potential out of existing rocket technology. Flight to the next level would now require a new generation of rockets.

Although von Braun and others of his team were based at Fort Bliss, they were often on the road, consulting with and advising contractors. Initially, they traveled with escorts, but as they demonstrated their reliability, they were allowed to travel by themselves. Of course, the Army kept von Braun and his men under loose surveillance, and on the occasions when von Braun lost his Army watchers, he was called upon to explain his whereabouts.[21]

The Federal Bureau of Investigation, which is responsible for ferreting out spies, subversives, and security risks inside the United States, also took notice of the Germans employed under Project Paperclip at Fort Bliss. On September 13, 1947, the director of the FBI, J. Edgar Hoover, wrote to the director of intelligence of the United States Army, Lt. Gen. Stephen J. Chamberlain, advising him that no classified technical information should be made available to the Germans. Hoover's advice was largely unenforceable and absurd because the Germans themselves had developed much of the classified information. Lieutenant General Chamberlain took the arrival of Hoover's uninvited advice as an opportunity to legitimize the presence of von Braun and his team in the United States. He advised Hoover that the Germans were allied in the battle against the threat presented by Communism. Hoover, who was opposed to Communism to the point of irrational obsession, accepted Chamberlain's arguments. He agreed to support the efforts of the Germans to get visas and become legal resident aliens.[22]

What Chamberlain did not tell Hoover was that at the very time he was pleading the case that von Braun and his team were not security risks, his own office was investigating allegations that von Braun, while vacationing in Germany, revealed American plans for the V-2 to a friend named Lewald.[23] In response to the charge against von Braun, his boss, Maj. James P. Hamill, admitted that von Braun was in Germany during the time in question and knew a man named Lewald, but he protected his own man with the conclusion that "It is the opinion of this office that the Dr. von Braun mentioned in reference letter and Professor Wernher von Braun of this suboffice are not the same person."[24] Clearly, the Army chose to protect von Braun, even though, by its surveillance and the content of its reports, it did not trust him.

The FBI opened a file on Wernher von Braun three months later, on August 16, 1948, for the purpose of evaluating "the internal security aspects of the immigration of von Braun into the United States for permanent residence." The FBI began its investigation by examining the dossier on von Braun compiled by the Joint Intelligence Objectives Agency (JIOA). While the document contained some embarrassing information, such as von Braun's memberships in Nazi organizations and the SS, it had been thoroughly purged of any suggestion that von Braun might be a security risk. Although a few interview reports carried negative comments on von Braun's character, the FBI ultimately concluded that there was no evidence that he was a security risk, and it had no objection to his legal immigration into the United States.[25]

On November 2, 1949, four years after he arrived in the United States, Wernher von Braun left Fort Bliss in the company of an Army escort. He crossed into Mexico from El Paso and went directly to the United States consulate in Ciudad Juarez, where he obtained a visa to enter the United States. A few hours after he left, Wernher von Braun returned to the United States as a legal immigrant.[26,27,28] By spring of the following year, most of the members of his rocket team "double crossed" the Rio Grande and became legal residents of the United States.[27]

One day, five years after arriving in the west Texas desert, as he walked with his colleague Dr. Adolf K. Thiel, Wernher von Braun brooded about his group's lack of progress, especially in the area of space exploration.

"We can dream about rockets and the Moon until Hell freezes over," von Braun said. "Unless the *people* understand it and the man who pays the bills is behind it, no dice. You worry about your damned calculations," von Braun said to Thiel, "and I'll talk to the people."[29]

In the years to come, von Braun would go over the heads of the generals and the bureaucrats and talk directly to the people who paid their salaries. He would sell them the moon.

5

Redstone

Today I live in a typical American town. Perhaps you have heard of it. The town is Huntsville, Ala., 80 miles north of Birmingham. Nearby are the vast establishments of the Redstone Arsenal of Army Ordnance. I am technical director of the arsenal's Guided Missile Development Group.

Here in America, even as an ex-enemy, I am free to go where I please and live where I please. I have just built my own three-bedroom cedar-shingle home with terrace on a nice hill at the edge of Huntsville.

I built this home with the help of an FHA loan. The monthly payment on it is $61, and that includes both mortgage payment and taxes. I think it is a good deal.

Wernher von Braun[1]

The defeat of Nazi Germany resulted in a brief intermission in the global dramas of the twentieth century. Three dominant ideologies were now down to two. The next act would pit the democracies of the West against Communism from the East. Winston Churchill recognized this division in 1946 when he declared that an Iron Curtain separated not only the two ideologies but also the Soviet Union and the countries under its influence from the rest of Europe. The ideological pushing and pulling that followed reached a crisis on June 24, 1948, when Joseph Stalin imposed a blockade on Berlin, which was an island of western democratic ideology on his side of the Iron Curtain. The Allies countered not with a military confrontation, but with an airlift to supply the isolated western sectors of Berlin, which were under their control. Several months later, in September 1948, the Soviet Union detonated its first atomic bomb, thereby substantially leveling any future battlefield.

Half-way around the globe, Communist forces took control of mainland China at the end of a protracted civil war. In October 1949, Chairman Mao Zedong proclaimed that the former ally of the western powers was now the Communist People's Republic of China.

If the United States was going to deal successfully with Communism, its leaders rationalized, it would once again need to be the militarily powerful and possess the most terrifying weapons on earth. On January 31, 1950, President Truman ordered the development of the hydrogen bomb. By then, the military services were already quietly planning rockets to deliver nuclear weapons. The Army's plans cleared a path for Wernher von Braun and his rocket team out of the desert and into the fertile land of Huntsville, Alabama.[2]

In 1805, John Hunt built his home near the natural spring at the foot of Monte Sano, on the northern side of the Tennessee River. His home became the nucleus of the town that eventually bore his name. Huntsville was the first English-speaking settlement in the area and the location for the constitutional convention of the region that in 1919 became the state of Alabama. The town prospered because of cheap land and the black slave labor that harvested cotton. The Civil War ended slavery, but cotton remained the premiere crop. Huntsville's economy was based on cotton, and the mills were the largest employers in the town, which boasted a population of 16,000 at the beginning of World War II.[3]

As involvement in the war against Germany appeared inevitable in 1941, the Army Chemical Warfare Service acquired about 30,000 acres of land at the southwestern edge of Huntsville, on the Tennessee River; this tract of land became the Huntsville Arsenal. Subsequently, Army Ordnance acquired an adjacent 4,000 acre parcel, which was ultimately named the Redstone Arsenal.[4] Huntsville became a boomtown as 20,000 employees at the Huntsville and Redstone Arsenals manufactured poison gas (which was never used in battle), and ammunition shells.[3,4] A portion of Huntsville Arsenal was built by German prisoners of war. They were young, defeated, and disillusioned. In 1944, about 1,100 prisoners of war were brought to the rolling green hills of northern Alabama, where they toiled under the direction of the United States Army. They built their own camp as rockets created by their fellow Germans blew apart buildings in London and Antwerp.

When peace came, munitions production was ended, Arsenal employees were laid off, and Huntsville Arsenal was shut down and scheduled for sale. In October 1948, the much smaller Redstone Arsenal, under the control of Army Ordnance, was designated by the chief of ordnance as the center for research and development of rockets.[4]

Holger Toftoy, now a major general and still head of the Army's rocket development program, arrived at the Redstone Arsenal on August 14, 1949. To his dismay, he found his future rocket development site totally inadequate. He learned, however, that the adjacent Huntsville Arsenal site was on the market, and he decided to go after it. He had a fight ahead of him because local business interests had their eyes on the Huntsville Arsenal as an asset for rejuvenating the local economy. Furthermore, Secretary of the Army Gordon Gray was opposed to transferring the property to Toftoy's rocket development program because he feared an implied commitment to greatly expanding the Army's rocket program. Toftoy took his case to the office of the army chief of staff. The only map he could find that showed both the Huntsville and Redstone Arsenals was larger than any wall available to him, so when Toftoy made his presentation, he rolled the map out on the floor and crawled over it on hands and knees to explain how he would use both sites. After he gave his pitch to Gen. Matthew B. Ridgway, who had the responsibility of making the final decision he said, "I'm really on my hands and knees, literally and figuratively begging for this place. Are there any questions?"[5]

Partly because of the Communist victory in China, the Army felt the need to expand weapons development, and Holger Toftoy got the much larger Huntsville Arsenal tract, which was assimilated into the Redstone Arsenal to form the Army's rocket development center.[5]

The migration from El Paso and White Sands to Huntsville began in April 1950, and was completed by November. Those personnel who moved included 130 German members of von Braun's team (there had been turnover in the original 118 man group, with some returning to Germany, others becoming employed by American companies, and additions of others from Germany under Project Paperclip), over 500 military personnel, 120 civilian government workers, and several hundred employees of the General Electric Company, the Army's prime rocket contractor.

The new organization was named the Ordnance Guided Missile Center. Holger Toftoy retained control and directed its operation from the rocket branch of the Ordnance Research and Development Division in the Pentagon. Col. James Hamill was his man in charge of rocket development at Redstone. Redstone Arsenal and its day-to-day operations were under the command of Gen. Thomas Vincent.[6] Wernher von Braun took the title of technical director, Guided Missile Development Group.[7] The structure of the organization at Redstone and von Braun's position in it were remarkably similar to Peenemuende during the Third Reich.

For Wernher von Braun and his group, the environs of Huntsville and northern Alabama, while geographically and culturally different from Germany, were more like home than the dry, dusty desert they had lived in for four years. For that matter, war and its aftermath insured that Germany was no longer the home they remembered and loved. Where Huntsville was lacking, then, the Germans would change it.

They enrolled their children in the public schools, began building homes, joined civic organizations, obtained library cards, and hiked through the wooded hills north of the Tennessee River. Huntsville did not have a Lutheran Church, so they organized a congregation, importing a minister from Florida.[8] However, von Braun, embracing the culture of his new home in America and, possibly more attuned to the social, political, and economic implications of church membership, joined an Episcopal congregation.[9]

As von Braun and the Army's rocket team were moving from their bases in El Paso and White Sands, North Korea, a country under a Communist government, gave the German team the biggest opportunity they had had since coming to America. On June 25, 1950, North Korean forces, supported by both the People's Republic of China and the Soviet Union, invaded South Korea, a nominally democratic country. President Truman sent in U.S. troops as part of a United Nations force to halt the invasion.[10] In July, the office of chief of Ordnance instructed the Redstone Arsenal group to conduct a feasibility study for a surface-to-surface ballistic missile with a range of 500 miles. As the war went on and von Braun's group made progress in the design, Army Ordnance raised the priority ranking of the new missile. Ordnance formally assigned responsibility for developing the missile to Redstone Arsenal on July 10, 1951. Ordnance also decided that it would be quite happy with a missile that had a range of 200 miles instead of the 500 originally requested.[6,11] As a bonus, the payload could be increased to three tons, enough to accomodate a nuclear warhead.[12] The new missile would have a range only slightly greater than that of the V-2, but if used in anger, its impact would make its predecessor look like a bottle-rocket in comparison. The rocket went through a series of names that included Ursa, Major, and then Redstone, its final designation on April 8, 1952.[6]

At about the time of his arrival in Huntsville, Wernher von Braun became reacquainted with the work of Robert H. Goddard, the American rocket pioneer whose work had encouraged Hermann Oberth and von Braun himself. Goddard's chief contributions were—along with Tsiolkovsky and Oberth—that he determined that the energy of liquid fuels was needed to lift a rocket into space and that he had built and launched the world's first liquid-fueled rocket in 1926. Between 1930 and 1941, with the financial backing of the Guggenheim Foundation, Goddard built many rockets and

launched them near Roswell, New Mexico, within range of a V-2 fired from White Sands to the southwest. Because some of his early work was ridiculed in the press, Goddard told the world little about it. He studiously patented every development, and the United States government, because of their military value, classified them. In 1941, with war approaching, Goddard became director of research for the Navy's Bureau of Aeronautics at the Naval Engineering Experiment Station at Annapolis, Maryland where he developed rockets to assist takeoff of Navy aircraft. Robert H. Goddard died in Baltimore on August 10, 1945, just as the Army negotiated with Wernher von Braun to bring his team to the United States.[13]

In 1950 James P. Hamill, now a lieutenant colonel, presented Wernher von Braun with a thick stack of papers and a problem. The papers consisted of more than 200 patents—still classified—that had been issued to Goddard. Goddard's estate, with the backing of the Guggenheim Foundation, sued the United States government for patent infringement. The estate claimed that the V-2s the Army was firing from White Sands and newer missiles designed and built in the United States infringed on Goddard's patents. Hamill asked von Braun to review the patents and write an assessment from an engineering standpoint, leaving out legal issues such as rights to war booty under international law.

Von Braun was possibly the first person working for the United States government to read the patents since the patent examiners had signed off on them. No one in Germany had seen them, and they were not the basis of any features of the V-2. However, there were, as von Braun discovered, infringements galore, even though von Braun's German team had invented the patented devices and features independent of Goddard. In his report, von Braun cited as possible infringements the use of jet vanes for flight path control, turbo-pumps to feed liquid fuels to the engines, gyroscopic guidance controls, and more.[14]

In 1960, a decade after Goddard's estate filed its suit, the government settled the claim. The Guggenheim Foundation and Goddard's widow split $1,000,000 and the U.S. government had the right to use the inventions described in over 200 patents. The settlement was probably inevitable since NASA had by then recognized Goddard's contributions by naming the Goddard Space Flight Center at Greenbelt, Maryland, after him on May 1, 1959.[13]

After reviewing Goddard's patents, however, von Braun concluded that Goddard had a brilliant and imaginative mind,[14] and he became an advocate of Goddard's memory. Von Braun's own record indicates that he had little interest in patents. Before entering the United States, he told interrogators that Germany had awarded him approximately ten patents dealing with rockets. These, he claimed, "had been taken over by the German Reich."[15] The patents remain obscure—if they exist at all. Von Braun's 1966 bibliography[16] records only one patent, "Rocket-Propelled Missile," which was issued to him by the United States Patent Office on January 21, 1961.[17]

Wernher von Braun's income as an employee of the United States Army allowed him to live as a comfortable member of the middle class. He did not have the perks and status he once had when he worked for Hitler's war machine, but he did not have the pressure to deliver instant victory either. He did have a new home financed with an FHA loan.[1] The von Braun family grew as Wernher and Maria's second daughter, Margrit Cecile, was born on May 8, 1952.[7] With two children who had United States citizenship and a career in America, von Braun was on the road to citizenship himself. He also gained a degree of personal independence when, during 1953, his parents returned to Germany where they lived on the pension his father had earned as a German government employee.[18]

No longer living in the desert, a legal resident at last, and having a respectable income—his 1951 contract with the Army paid him $10,500 for the year, good money for the time—[19] von Braun now made time for leisure activities. He took up boating and water skiing on the Tennessee River, earned a license to pilot small aircraft and in the summer of 1954, at the urging of science fiction writer Arthur C. Clarke, he took up scuba diving.[20]

As his home life became tranquil, von Braun acquired professional recognition and respectability. He was now, after all, on the team that defended the Free World against Communism as well as a leading theoretician on space travel. For the latter he was honored with memberships in the British Interplanetary Society, the American Rocket Society, and the Explorer's Club (New York).[21] In September 1951, the British Interplanetary Society hosted the Second International Congress of Astronautics, which had as its official subject, the "earth satellite vehicle." Sixty-three delegates from ten countries came prepared to discuss the feasibility of putting the first artificial satellite into orbit around earth. Arthur C. Clarke, chairman of the society, pompously asserted that "Space exploration is likely to be the next major technical achievement of our species." Wernher von Braun did not attend the conference but his paper to the congress was read for him. He took the artificial satellite as a given, and presented a plan for an expedition to Mars, taken from his unpublished book *The Mars Project*, that contained enough quantitative detail to give it credibility. While everyone else was trying to figure out how to put a tin can into orbit, von Braun told them how to go to Mars, explore its surface, and return safely. Without leaving his Huntsville home, von Braun staked his claim as the leading visionary in the small community of space enthusiasts.[22]

The following year, in 1952, von Braun was visited in Huntsville by Richard Bechtle, a German publisher. Von Braun showed him his manuscript for *The Mars Project*, and Bechtle agreed to publish the scientific appendix as a small, technical monograph. *Das Mars Projekt* appeared in

1952.[23] The University of Illinois Press picked up the English rights and published *The Mars Project* in 1953.[24] Bechtle also talked von Braun into taking on a professional writer, a man named Neher, who would turn it into a science fiction novel and give it the zip it needed to be commercially viable. Neher took so many liberties with the laws of physics that von Braun withdrew as coauthor.[25,26]

In addition to the partial success of *The Mars Project, Collier's* magazine invited him to participate in writing a series of articles on space exploration.

Von Braun's family life was serene. His career was, it seemed, back on track as technical director of the Army's rocket development program at Redstone Arsenal. He was beginning to get his plans for space travel to the people. As Wernher von Braun reached the age of forty in 1952, seven years after his arrival in the custody of the United States Army, he had reason to believe that life in America was good.

6

Selling Space: The *Collier's* Articles

> This man [Wernher von Braun] could convince anybody. His dreams,
> his ideas are mesmerizing. He is so effective that he could sell anybody
> anything. Even used cars!
>
> —Cornelius Ryan[1]

At the midpoint of the twentieth century, outer space was owned by writers
of science fiction. Space travel was the pursuit of Flash Gordon, Buck Rog-
ers, and the legions of space warriors who followed them. Science took its
first steps toward space in the 1920s and early 1930s through the inspiration of Robert Goddard and Hermann Oberth, then lost the battle to
fiction as the world stumbled into the Second World War. Ironically, one
product of that war, the V-2 rocket, reignited interest in the scientific approach to space. Wernher von Braun and his team showed that big rockets
could reach into space, at least briefly. The Army's Bumper project demonstrated that a second stage boosted by the V-2 could reach hundreds of
miles into space. It wasn't much of an extrapolation then to imagine a
larger rocket with additional stages carrying large objects and even men
into space. Mankind was ready to step into space; it only needed a plan.

Serious scientific planning for the exploration of space began with the

Space Travel Symposium held at the Hayden Planetarium in New York City on Columbus Day, October 12, 1951.[2]

Willy Ley, a writer on popular science subjects in general and on rockets in particular, coordinated the Symposium. Along with Wernher von Braun, Ley was a member of the *Verein fur Raumschiffahrt* (VfR) and the *Raketenflugplatz* group in Germany during the late 1920s and early 1930s. He knew von Braun, but they apparently were not close friends. After von Braun went to work for the German Army and after the Nazis came to power, Ley went on an extended vacation that ultimately brought him to the United States; he never returned to Germany.[3]

Invitations to the Space Travel Symposium were sent to academic institutions, research groups, and professional societies: people who took space travel seriously and could make it happen. The news media were also invited. At 9:00 A.M. on Friday morning, October 12, over 200 people were in attendance at the Hayden Planetarium including Arthur C. Clarke and many members of the armed forces, but not Wernher von Braun. Two writers from *Collier's* magazine, one of America's most popular publications at the time, also attended.[4,5] A half-dozen symposium speakers discussed topics relating to space travel that ranged from "Engineering and Applications of the Space Vehicle" (R. P. Haviland) through "Space Medicine" (Heinz Haber) and "Legal Aspects of Space Travel" (Oscar Schacter).[2]

The *Collier's* writers reported back to their editor that there was a story in space travel, and that the magazine should cover it. Cornelius Ryan, who joined *Collier's* staff as an associate editor just that year, was delegated to follow up on the story by attending a space medicine conference to be held from November 6 to 9 in San Antonio, Texas. (Ryan would later write the best-seller, *The Longest Day*, an account of the Normandy invasion, and become editor of *Collier's*.) The symposium was sponsored jointly by the Lovelace Foundation for Medical Education and Research of Albuquerque, New Mexico, and the United States Air Force School of Aviation Medicine. Wernher von Braun was invited to attend the symposium, but was not scheduled to make a presentation.

Cornelius Ryan showed up in San Antonio with limited interest in space travel and a healthy skepticism about the subject. After one day of presentations, Ryan found himself cornered in a dining room by Fred Whipple, chair of the Department of Astronomy at Harvard University, Joseph Kaplan, professor of Physics at UCLA, and Wernher von Braun. They all knew what kind of publicity Ryan could secure for them on their favorite topic of space travel, and they went to work on the trapped writer. They fed him cocktails and arguments, dinner and explanations, more drinks, and more answers. By the time the three men were done, it was midnight, Ryan was exhausted, but a new believer in space exploration.[2]

Ryan returned to his office in New York to plan *Collier's* coverage of space travel. He began by organizing an in-house symposium at *Collier's* that featured the experts who participated in the Hayden Planetarium and San Antonio meetings and the best salesman of the group, Wernher von Braun. He also enlisted three top illustrators to turn the ideas of the experts into images to catch the attention of *Collier's* readers. It quickly became clear that the material justified several articles. In the end, *Collier's* published eight articles over a two-year period.[5]

Wernher von Braun spent twenty-five years developing rockets and pioneering astronautics, but for the first twenty of those years he and his ideas were kept under tight control, first by the German Army, then by the United States Army. Now, because he would only be speculating as part of a panel of experts about the distant future, he was permitted to place himself and his ideas about space travel before a broad audience. For him, writing for *Collier's* was play rather than work.

> After a day of excruciating meetings for the Redstone Project . . . , it is such an enjoyable relaxation to transpose yourself to the lunar surface and simply charge ahead with a colorful description of all the exciting adventures that expect you there. . . . I mix some martinis, put a Brandenburg concerto on the record player; and just write and write . . . until Maria gets out of bed and reminds me that I must be in the office in two hours from now.[6]

Writing for *Collier's* was doubly sweet for von Braun by the fee he received, $1,000 for a 5,000 word article,[7] a welcome bonus for a civil servant who was paying a mortgage, raising a family, and supporting his elderly parents on $10,500 per year.[8]

The experts returned their completed articles to *Collier's* by the end of 1951, and then Ryan, with the help of Willy Ley, did some heavy editing to make their writing fit *Collier's* style and be intelligible to a lay audience. Ryan also carried them to Washington to get clearance because two of his writers, Wernher von Braun and Heinz Haber were employees of the federal government.[9]

The first articles on space travel appeared in *Collier's* March 22, 1952, issue that arrived on newsstands about one week earlier than the issue date. Concurrent with the issue's release, the publisher organized a publicity blitz with Wernher von Braun as its point man. He was scheduled for seven television interviews with the top media people of the day: John Cameron Swayze, Dave Garroway, and Garry Moore.[10] The magazine claimed a circulation of over three million copies and that each copy was read by four or five people.[5] While *Collier's* may have exaggerated their readership somewhat, the base circulation of the magazine and the television exposure put von Braun on his way to becoming a nationally recognized, if not influential, figure.

The cover of the March 22, 1952, *Collier's*, showed a dramatic, above-the-earth view of the third and final stage of a man-carrying rocket separating from its second stage as it accelerated into orbit. The text on the cover read, "Man Will Conquer Space *Soon*. Top Scientists Tell How in 15 Startling Pages."[11] On page 23 the contributors to the articles were introduced: Dr. Wernher von Braun, technical director of the Army Ordnance Guided Missiles Group; Dr. Fred L. Whipple, chair, department of Astronomy, Harvard University; Dr. Joseph Kaplan, Professor of Physics at UCLA; Dr. Heinz Haber, the U.S. Air Force department of Space Medicine; Willy Ley, at the time, probably the best-known magazine science writer in the United States; Oscar Schachter, deputy director of the UN Legal Department; and Chesley Bonestell, Fred Freeman, and Rolf Klep, artists and illustrators of the articles. The contributors' credentials made it clear that *Collier's* wanted their articles to be taken seriously as fact, not fiction.[12]

The introductory comment, "What Are We Waiting For?" set forth the priorities and tone that guided the series of articles and the ventures into space to follow.

> On the following pages, *Collier's* presents what may be one of the most important scientific symposia ever published by a national magazine. It is the story of the inevitability of man's conquest of space.
>
> What you will read here is not science fiction. It is serious fact. Moreover it is an urgent warning that the U.S. must immediately embark on a long-range development program to secure for the West "space superiority." If we do not, somebody else will. That somebody else very probably would be the Soviet Union.[12]

Most of the rest of the introductory editorial continued with the same theme: The United States must establish its presence in space before the Soviet Union does in order to establish its military superiority and to insure its security. Space policy was not developed in a vacuum but as part of the perceived needs and realities of the Cold War.

The article by Wernher von Braun, "Crossing the Last Frontier," was the centerpiece of several articles on space in the issue.[13] The article made it clear that its author had a well-conceived agenda: "Within the next 10 or 15 years, the earth will have a new companion in the skies, a man-made satellite that could be either the greatest force for peace ever devised or one of the most terrible weapons of war—depending on who makes and controls it." Fortunately, von Braun quickly left behind the international politics of the day and got into the business of rockets, satellites, and space travel.[13] While his proposals may have been revolutionary to *Collier's* readers, they were nothing new to von Braun. He had presented the basic concepts and many of the details to the El Paso Rotary Club in his first public appearance in America five years earlier.[14]

The venture into space begins with a rocket, bigger and more powerful than any imagined before. It is composed of three stages, is 265 feet high, and has a total weight of 7,000 tons. The first stage has a cluster of fifty-one engines, which when fired simultaneously, produce a total thrust of 14,000 tons. (By comparison, the Saturn V rocket that would carry astronauts to the moon would have a relatively smaller take-off weight of 3,000 tons and thrust of 3,750 tons. Yet, because of improved technology, it would be able to lift 125 metric tons into orbit.)[15] The first and second stages, after burning out, are parachuted to the ocean and retrieved for reuse. The third and final stage carries a payload of thirty-six tons of cargo and crew. It reaches an orbit of 1,075 miles and circles the earth every two hours. The third stage is a winged vehicle, and after it's mission is completed, it glides back to earth on these wings much like the space shuttle of the late twentieth century.[13]

Multiple rockets bring more space men and the components of the space station into orbit. The prefabricated components are made of nylon and plastic, and they are assembled by men working in the vacuum and weightlessness of space. When the components are assembled, air is pumped into the space station, inflating it to its designed shape.

The space station is shaped like a tire, 250 feet in diameter, and like a tire it rotates around its hub once every 12.3 seconds. The purpose of the rotation is to create centrifugal force in the areas away from the hub, an artificial gravity in which the satellite's crew can work in comfort. (By 1952, most experts in space medicine discount the possibility of weightlessness posing a serious danger to humans.)[13] The concept of the wheel shape to create artificial gravity was originally conceived by an Austrian army captain named Potocnik writing under the pseudonym of Hermann Noordung.[16] Von Braun, who exploited good ideas wherever he found them, incorporated it in his plans for his space station.

The space station, as imagined by Wernher von Braun, is accompanied in space by a remote-controlled, free-flying telescope. It conducts astronomical research and is similar to the Hubble space telescope, which is put into orbit decades later.

In his *Collier's* article, von Braun presented a brief description of how the space station could be used as a staging area for deeper penetration of space, specifically a voyage around the moon and back. The way he described it, it was a simple job of cobbling together a suitable vehicle to house the crew for a ten-day trip, pointing it in the right direction, and firing a rocket motor for two minutes.[13]

In addition to being a staging area for further space exploration, von Braun's space station is used for meteorological observation, military reconnaissance, and as a nuclear battle station. Von Braun writes:

> Small winged rocket missiles with atomic warheads could be launched from the station in such a manner that they would strike their targets at supersonic speeds. By simultaneous radar tracking of both missile

and target, these atomic-headed rockets could be accurately guided to any spot on the earth.

In view of the station's ability to pass over all inhabited regions of earth, such atom-bombing techniques would offer the satellite's builders the most important tactical and strategic advance in military history.[13]

In the thinking of 1952, when national survival was at stake, von Braun's proposal was totally reasonable. If push came to shove, the United States could destroy Moscow the way it had incinerated Hiroshima and Nagasaki, but instead of dropping a bomb from an airplane six miles high, it could be done antiseptically and impersonally from over a thousand miles. The Communists would do it to the United States if they could, wouldn't they?

Other articles in the March 22, 1952, issue of *Collier's* fleshed out the space theme. "A Station in Space" by Willy Ley gave the fine details of the space station's design and life within it.[17] In "The Heavens Open," Fred Whipple explained the advances in astronomy that could be made from an observatory in space.[18] Joseph Kaplan in "This Side of Infinity" briefly explained what was known about the space between the surface of the earth and outer space, where the space station soared.[19] "Can We Survive in Space?" by Heinz Haber examined the question and came up with the answer: yes.[20] In his article, Oscar Schachter asked, "Who Owns the Universe?" and answered that this legal question is still open.[21] "Space Quiz, Around the Editor's Desk" was a question and answer section that filled in details and summed up. Wernher von Braun got the first and last words in this piece. It ends with this exchange:

Q. Would Soviet Russia enjoy any advantages in a race for space superiority?

Von Braun: Just one advantage of any importance so far as is known. Because the country is huge and barricaded behind the Iron Curtain, the initial phases of a space program could be kept secret much more easily in the Soviet Union than in the western World. . . . the advantage in the competition to conquer space probably rests with us—if we move quickly.[22]

What made the *Collier's* articles—and especially von Braun's—so compelling and believable, was the quantitative data presented by the authors. A fundamental feature of science is calculation, and the authors' numbers pushed their speculations in the direction of scientific fact. The numbers derived from the laws of physics were, of course, virtually impossible to challenge. Von Braun's cost figures were clearly based more on speculation than science. He estimated that the entire cost of establishing a space station, including development, constructing a fleet of rockets, and fuel factories would be about $4 billion. When production of spacecraft became routine, he estimated that the cost of each—he did not indicate if he meant the entire vehicle or just the third stage—would drop to about $1 million, about the cost of a large airliner in the early 1950s.[22]

With the publication of its March 22, 1952 issue, *Collier's* staked its claim as the authoritative source of up-to-date information on man—or humans,—in space. Apparently, many readers of *Collier's* enjoyed the first issue on space and became avid enthusiasts of the subject. *Collier's* prepared for a series of space issues.

Collier's second space issue, dated October 18, 1952, featured on its cover an artistic interpretation of a space vehicle—lacking any aerodynamic features—landing on the surface of the moon with the earth hanging in the sky in the background. The cover text reads, "Man on the Moon: Scientists Tell How We Can Land There In Our Lifetime."[23] The centerpiece of the issue is a lavishly illustrated article by von Braun titled "Man on the Moon: The Journey."[24] Two other brief articles rounded out the theme.

Von Braun's second article begins, "Here is how we shall go to the moon." The sentence is strangely reminiscent of the line from T. S. Eliot's poem "The Hollow Men": "This is the way the world ends, Not with a bang but a whimper."[25] Was this line rolling around in von Braun's subconscious as he wrote his article? Was von Braun its author, or did Cornelius Ryan sneak it in as a rebuttal to the cynical poet and as an affirmation of resolve and hope? The first sentence is a statement of total confidence and commitment; there is no hint of a conditional *if* or *when*. And who, reading these words in 1952 when the trip is still a dream, would not want to go on the journey to the moon?

The expedition to the moon begins 1,075 miles above the earth, in orbit near the space station. Rockets from earth ferry up prefabricated components that are to be assembled into three moonships. Each vehicle is bulky: 160 feet long and 120 feet wide. They are open scaffoldings on which are hung crew compartments, fuel tanks, and rocket engines. Two of the moonships house twenty crew members each. The third vehicle is a cargo ship that houses only ten crew members, but has a large supply and cargo hold. The moonships are not streamlined; there is no need for it because the vessels will not encounter any friction-causing atmosphere on their journey.[26]

The journey begins when the rocket engines of the moonships are fired for a matter of minutes to increase their speed relative to the surface of the earth by 3,660 miles per hour. (The moonships, like the space station, are orbiting the earth at 15,840 miles per hour.) This relatively small increase in speed is enough to bump the ships out of earth orbit and propel them in the direction of the moon.

After five days of travel, the moonships are near their destination. An automated system fires the rockets that gently lower the three vehicles to the surface of the moon; landing a moonship is considered too "tricky" to trust to human pilots. The spider-like legs of the moonships strike the soft, volcanic ground; the lunar expedition has landed. Exploration can begin.[24]

In reading von Braun's article, it is interesting to note the differences between his 1952 prediction and the reality of *Apollo 11* seventeen years later. First, an introductory piece, "Man on the Moon"[26] stated, "Our trip to the moon will not be a simple nonstop flight from the earth. We'd need too large and expensive a rocket ship for that." In fact, a multistage rocket to the moon and back was chosen as the vehicle because it was the cheapest and simplest scenario. Second, the Apollo expeditions used relatively small lunar modules with only two crew members. Besides minimizing the cost, this approach also lowered the risk to personnel. And third, the Apollo missions had pilot-astronauts guide the lunar modules to the moon's surface. Their quick reactions and judgment were needed to guide their vehicles over small craters and dangerous boulders.

A week after the "Man on the Moon" issue of *Collier's* appeared, the October 25, 1952, issue contained von Braun's article "The Exploration of the Moon."[27]

On landing, the lunar mission crews unload supplies and equipment from the cargo vehicle. The cargo includes three vehicles for exploration of the moon's surface at distances far from the landing site. The cargo hold itself is constructed as two longitudinal parts which, when emptied of cargo, are separated and buried. One half becomes a laboratory, the other, living quarters. The lunar soil above them provides protection from meteorites and radiation. After six weeks of exploration of the moon's surface, the crews prepare for their return to earth. Twenty-five crew members board each of the two passenger ships for the return flight. The cargo ship and all it carried are left behind.[27]

The first three issues of *Collier's* space series were greeted enthusiastically not only by its subscribers, but also by Paramount Pictures Corporation in Hollywood. By the end of November 1952, the studio had the script for a space movie that von Braun judged as "sound and suitable as a basis for further discussion."[28] Von Braun was in contact with Paramount's top management, and he intended to bring along Willy Ley as technical consultant on the movie project.[28] However, Von Braun's first attempt at a movie project did not work out. *Collier's* owned the copyrights to the space exploration articles, and it intended to sell all rights, including reprints and movie rights, as a package. Paramount apparently lost interest, and a movie based on the *Collier's* articles was never made by Paramount or any other studio.[29]

After publication of its first three space issues, *Collier's* was flooded with inquiries about the human aspects of space exploration. Readers wanted to know how the spaceship crews would be selected and trained. *Collier's* responded with three more issues that addressed these questions under the broad but unfortunate title, "Man's Survival in Space." All those who participated in the original symposium were given credit as contributors, though Cornelius Ryan probably did the lion's share of the writing.

The February 28, 1953, issue of *Collier's* carried an article titled "Picking the Men." It conceded that women will "be sought after for some space crew jobs," although not as pilots.[30] The article gave a lot of space, including its cover photograph, to a high-altitude pressure suit developed for the Navy.[30] The March 7, 1953, issue had an article that succinctly stated its content: "Testing the Men." After lengthy descriptions of the testing regimens, there is a section on "Reasons for Ban on Women." It stated that "Women, who may beat out men for certain crew jobs, won't go along on interplanetary journeys, where privacy will be lacking for long periods. So they'll take the . . . tests separately, and briefly, in preparation for the shorter flights that they will make."[31] Social and cultural beliefs were an integral part of the planning.

The March 14, 1953, issue had an article titled "EMERGENCY!" giving drama to an already dramatic subject. It asked and answered the question, "What happens when disaster strikes in space? Can the crew of a 15,000-mile-an-hour rocket ship bail out or land their disabled craft?" The affirmative answer engenders a feeling of frustration and a sense of pointless loss to everyone who watched the *Challenger* space shuttle disintegrate thirty-three years later. The authors of the article, led by Wernher von Braun, described individual ejection pods for each crew member that bring them by parachute safely to earth.[32] The pods might not have worked under all circumstances, but they might have given the *Challenger* crew a chance to survive.[32]

The seventh issue of *Collier's* on the space theme had a cover illustration of a cone-shaped satellite in orbit with New York below it to the left and Cape Cod, Massachusetts, below it to the right. "The Baby Space Station," the cover blurb said, "First Step in the Conquest of Space." The article inside was bylined: by Dr. Wernher von Braun with Cornelius Ryan. The project described in the article could be the most ambitious one compared to those that preceded it. Braun described a project that is just over the technological horizon and could actually be accomplished without the Herculean effort needed for the big space station project.

Von Braun's article describes a small satellite that circles the earth at an altitude of 200 miles and completes one orbit every ninety-one minutes. The satellite collects data on cosmic radiation, meteoric dust, and weightlessness. Its passengers are three rhesus monkeys, man's surrogates on this first prolonged stay in space. After about sixty days in orbit, friction with the upper atmosphere slows the satellite until it falls from orbit. The monkeys get a dose of lethal gas just before the satellite burns up on reentry.

The baby space station could be put into orbit in the relatively short term and for relatively little money, yet would have a big technical payback. Von Braun saw it as a first step in the space station project, which, he predicted, could be completed in five to seven years. He is ready to start.[33]

The eighth and final issue that *Collier's* devoted to space travel was published on April 30, 1954, and it presented an inspired conclusion to the series. The cover illustration showed eight spaceships, most of them similar in construction to the moonship of eighteen months earlier, as they approached the red planet, seen in the background. Above the illustration are the questions: "Can We Get to Mars?" and "Is There Life on Mars?"[34]

Fred Whipple addressed the question of life on Mars first in a half page article. He concluded his analysis with, "There's only one way to find out for sure what is on Mars—and that's to go there."[35]

"Can We Get to Mars?" by Wernher von Braun with Cornelius Ryan concluded that a voyage to Mars is inevitable, but the author predicted that it would not take place for a century or more.[36] Then why plan the trip now? Why not? It is a daring goal, it fires the imagination, it drives future thinking about space, and it sells magazines.

Von Braun had described the overall scenario of a Martian expedition and mathematically demonstrated its possibility in *The Mars Project*,[37] which was published in Germany in 1952 and in 1953 by the University of Illinois Press. As with the previous articles, the collaboration of Ryan, as editor, and superb artwork make the *Collier's* article about an expedition to Mars accessible and fascinating to lay readers.

For the mission to Mars, a fleet of ten spaceships, each weighing about 4,000 tons, is assembled in orbit around the earth. Seven of these ships are primarily cargo and housing facilities; they carry the supplies and equipment for the seventy men on the trip. Three are landing craft that will undergo final assembly on Mars. The journey begins as the ships fire their rockets, increasing their speed enough to escape the earth's orbit and to put them on a trajectory that will take them the 355,000,000 miles to Mars.[36]

After eight months in transit, the spacecraft again fire their rocket engines, this time to decrease their speed and put them into orbit 600 miles above the surface of Mars. The explorers then assemble their first landing craft, which appears to be a huge glider with swept-back wings and a width five times its length. It is believed, in 1954, that the Martian atmosphere is thick enough to support an aircraft of this design. An advance party flies the first vehicle to a landing site on the planet's snow-covered polar ice cap. This is the only area on Mars where the explorers have reason to believe the surface is smooth enough to be used as a landing field. The crew then unloads its supplies and tractors, which they will drive 4,000 miles to the equator. The advance party then constructs a landing strip, which is used by the remaining members of the crew to descend to the Martian surface in two more winged craft with additional supplies.

The seventy-man crew spends fifteen months exploring the Martian surface and conducting experiments. When the positions of the earth and Mars in their orbits make a return trip feasible, the two landing craft are prepared for take-off. Their wings, which are now dead weight, are stripped off, and the fuselages tilted to stand on their tails. Rocket engines lift the two craft and their crews back into orbit around Mars, then back home. The vehicle that landed on the polar ice cap and most of the equipment are left behind.[36]

After their publication in *Collier's*, the articles by Wernher von Braun appeared again in expanded, book form. *Across the Space Frontier*, which was based on the first *Collier's* articles and edited by Cornelius Ryan, arrived in bookstores in September 1952.[38] *Conquest of the Moon*, which was based on the two lunar exploration issues, was also edited by Cornelius Ryan and was published one year later.[39] Willy Ley and Wernher von Braun expanded the last articles of the series into *The Exploration of Mars*, which appeared in 1956.[40] With the publication of the *Collier's* series and republication in more durable book form, Wernher von Braun became the ranking authority on space exploration in the minds of the public. Yet, as von Braun understood, to plan space travel is very different from actually having the resources to do it.

In addition to working on the *Collier's* articles, Wernher von Braun was still on the Army's payroll as chief of the Guided Missile Development Division at Redstone Arsenal,[41] and he earned his keep. During this time, his group developed the Redstone ballistic missile. While it was often described as a second generation V-2, the Redstone incorporated many advances in missile design. It was made of aluminum. Its fuel tanks were of the monocoque type: they were an integral part of missile's structure and their walls were the outer walls of the rocket. The Redstone missile had an all new, completely internal guidance system. And, of course, it could carry a nuclear warhead.[42,43]

The first Redstone was fired from Cape Canaveral on August 20, 1953. Many years would pass, however, before it completed its testing and became operational. The Redstone was simple in comparison with its support requirements. The missile was only one piece of hardware in the hands of a 600-man Field Artillery Missile Group that also had to contend with numerous large vehicles and a mobile, liquid-oxygen production plant.[42]

Germany's rocket pioneers in 1930. From left to right: Rudolf Nebel, Dr. Alexander Ritter of the Chemisch-Technische Reichsanstalt, Hans Bermueller, Kurt Heinisch, Hermann Oberth at the right of a rocket of his design, two unidentified men behind Oberth, Klaus Riedel holding a "Mirak" rocket, Wernher von Braun, and another unidentified man. Photograph courtesy of NASA, Marshall Space Flight Center.

Wernher von Braun (right front), despite having his arm in a cast, was confident and in good spirits after he, Gen. Walter Dornberger (left front, holding a cigar), and others surrendered to the United States Army on May 2, 1945. Photograph courtesy of the National Archives, photo no. 208-PU-212KK-1.

By early 1946, Project Paperclip had brought 118 German rocket builders to the United States to continue their work. In this group photograph taken at White Sands, New Mexico, Wernher von Braun is seventh from the right in the front row; Arthur Rudolph, production manager of the Mittelwerk and later director of Saturn V production, is fourth from the left in the front row. Photograph courtesy of NASA, Marshall Space Flight Center, Public Affairs Office.

A V-2 first stage with a WAC/Corporal second stage lifts off from the White Sands Proving Grounds as part of the Army's Bumper program. Photograph courtesy of NASA, photo no. 67-H-1452.

Wernher von Braun describes the "bottle suit" concept he developed to Heinz Haber and Willy Ley, his fellow advisors on Walt Disney's "Man in Space" television program. Photograph courtesy of NASA, Marshall Space Flight Center.

Wernher von Braun appeared in the Defense Department's informational and educational film *The Challenge of Outer Space*, which was released February 29, 1956. He is seen here holding a model of his proposed three-stage orbital rocket and, behind him, illustrations of the winged orbital vehicle and the wheel-shaped space station. Photograph courtesy of the Smithsonian Institution, Neg. No. 77–12796.

Wernher von Braun poses with some of the key staff of the Army Ballistic Missile Agency (ABMA) at Huntsville, Alabama. From left to right: Ernst Stuhlinger (seated); Maj. Gen. Holger N. Toftoy, commanding officer; Hermann Oberth (foreground), the father of German rocketry who worked for the Army Ballistic Missile Agency at Huntsville from 1955 to 1959; Wernher von Braun; and Eberhard Rees, deputy director, Development Operations Division. Photograph courtesy of NASA, photo no. CC-417.

Sergei Korolyov, survivor of Stalin's gulag and the chief designer of the Soviet Union's rocket program, was Wernher von Braun's anonymous foil and nemesis. Photograph courtesy of the Smithsonian Institution, Neg. No. 76–17276.

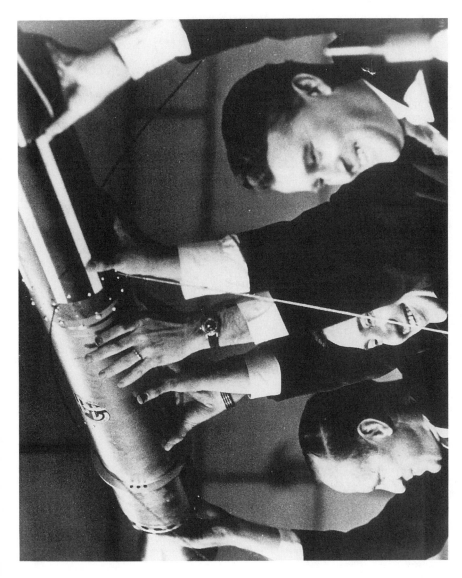

At a press conference following the successful launch of the United States' first satellite, William H. Pickering, director of the Jet Propulsion Laboratory, James A. Van Allen, head of the team responsible for instrumentation, and Wernher von Braun triumphantly held aloft a replica of *Explorer I.* Photograph courtesy of the U.S. Army Missile Command Historical Office.

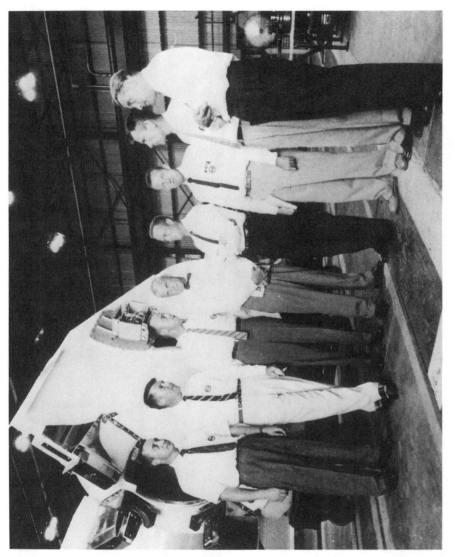

Before going into space, America's first seven astronauts met with Wernher von Braun in 1959 at the Army Ballistic Missile Agency (ABMA) in Huntsville, Alabama. Pictured from left to right: Gus Grissom, Wally Schirra, Alan Shepard, John Glenn, Scott Carpenter, Gordon Cooper, Deke Slayton, and Wernher von Braun. Photograph courtesy of NASA, Marshall Space Flight Center.

Wernher von Braun explained the Saturn rocket to President Dwight Eisenhower when the latter spoke at the dedication ceremony of the Marshall Space Flight Center in Huntsville, Alabama, on September 8, 1960. Photograph courtesy of NASA, Marshall Space Flight Center.

The Mercury-Redstone III carries Alan Shepard, America's first man in space from Cape Canaveral on May 5, 1961. Photograph courtesy of NASA, photo no. 61-MR3-72A.

Wernher von Braun posed in his office with models of his rockets. An early concept drawing of a lunar landing vehicle is in the background. Photograph courtesy of NASA, Marshall Space Flight Center.

7

Disneyland

> Now, when we opened Disneyland, outer space was Buck Rogers. I did put in a trip to the moon, and I got Wernher von Braun to help me plan the thing. And of course we were going to the moon long before Sputnik. And since then has come Sputnik and then has come our great program in outer space.
>
> Walt Disney[1]

In the early 1950s, Walt Disney, who made his name and fortune by elevating animated movie cartoons to a feature length art form, turned his attention in the direction of what he called a leisure park. He wanted to create a clean, wholesome place where both children and their parents could have fun, a place with much of the character of Tivoli Gardens in Copenhagen, which greatly impressed him.

Walt's brother and partner Roy, who ran the business affairs of the Walt Disney Company, thought the leisure park idea would lead to disaster and refused to let the company get financially involved. Walt, confident of his concept, set up W.E.D. Enterprises (for Walter Elias Disney) to create his park. He put most of his own savings into the initial planning but failed to interest the investors he needed to bring Disneyland into being.[2]

Disney then came up with the idea of working with the new and growing television networks to finance Disneyland. At the time, by tacit agreement, no studio would license its movies for broadcast on television. Disney proposed to create a series of programs made specifically for television, and not theater release, in return for financial backing for his leisure park. Walt's brother Roy liked the idea and the brothers finally began working together on the project. In the spring of 1953, Disney reached an agreement with the American Broadcasting Company, then the smallest of the three television networks, in which ABC would put up $250,000 in cash and guarantee loans up to $4,500,000. Construction of Disneyland began in the summer of 1954 when bulldozers began tearing out orange trees from a 160 acre parcel of land in Anaheim, California.[3]

In wanting a partnership with a television network, Disney had no thoughts of bringing art to the new media, but he was aware of television's great advertising power. One of Disney's first projects for ABC, therefore, was an hour-long program broadcast on Sunday evenings called *Disneyland*. It was a one-hour, prime-time advertisement for the park of the same name and for other Disney projects, and since the television network was an investor in the park, it was thrilled to have the program. The premiere showing of *Disneyland* on October 27, 1954, described coming attractions at the park in Anaheim and on the television program itself.[4] The second *Disneyland* program was about the making of *20,000 Leagues Under the Sea*, a Disney film that was about to appear in theaters.[5]

Walt Disney created areas in his leisure park called Fantasyland, Frontierland, Adventureland, and Tomorrowland. The first three leaned heavily on characters and themes from Disney movie productions. Tomorrowland, however, was unexplored territory, not based on anything the Disney studio worked on before. Walt and W.E.D. Enterprises started with a clean sheet of paper. Disney asked one of his top animators and producers, Ward Kimball, to develop a television program that covered Tomorrowland and, not incidentally, to come up with ideas for the park. Kimball had worked for Disney since 1934 on the now classic animated movies, *Snow White and the Seven Dwarfs*, *Pinocchio*, and *Fantasia*.[6] Kimball was no futurist, but he had been following the *Collier's* series on space, and he thought that rockets and the exploration of space would probably deliver a vision of tomorrow that might capture Disney's imagination. On April 17, 1954, Kimball showed Disney his sketches and screen treatment for a television film with the theme of space exploration. Disney was ecstatic. The characteristically tight-fisted Disney, symbolically handed Kimball a blank sheet of paper and said, "Write your own ticket!"[7] Before it was over, Kimball produced three one-hour programs at a total cost of one million dollars, an unprecedented sum of money to spend on television programs in the 1950s.[6]

Disney demanded that his Tomorrowland program be based on scientific fact, not science fiction. Extrapolation of technical fact was fine, but fantasy had already been given a land of its own. Kimball, with copies of back issues of *Collier's* in hand, knew where he could get the technical help he needed. First he called in Willy Ley. With his encyclopedic knowledge and entertaining style, Ley made a strong impression on Kimball and his staff. Kimball then invited Heinz Haber, the physiologist, and Wernher von Braun, the rocket designer, to be consultants. Both accepted and Kimball had assembled his well-rounded team.

Geography was no problem for von Braun because he constantly traveled from his home in Huntsville, Alabama, to Los Angeles to consult with contractors on the Redstone and Jupiter rockets. He conducted Army business during the day, then went to the Disney studio when he was free to work the night shift.[7] According to one account, Disney assigned to the project a top-salaried, very attractive woman artist on his staff to brew coffee, serve pastries, and keep the tape machine playing classical music. It seems to have been an idyllic work environment except, possibly, for the woman artist who was assigned to domestic chores.[8]

Von Braun took both a creative and quantitative approach to giving the scientific facts that Disney demanded. He sketched out designs for spaceships and satellites. He calculated the dimensions of vehicles and satellites and the parameters of space flight. Disney wanted accurate models of the spacecraft and the satellites, and von Braun kept Disney's model builders in touch with reality by providing technical information that ranged from in-orbit fueling operations to the preparation of meals under weightless conditions.[7]

Working sessions often went late into the night. At the end of one long day of work at the Disney studio, one of the writers assigned to the Tomorrowland project, Charles Shows, offered to drop von Braun at his hotel on his way home. Von Braun accepted, and they were off, talking as they drove. As usual, von Braun did most of the talking: about his interest in space travel, his work for the Nazis during the war, and his life in the United States. Shows asked von Braun if there was anything about America that he did not like.

"Yes," von Braun said bluntly. "I don't like being treated like a foreign spy. Everywhere I go, the FBI has me followed. I can't even go to the bathroom without an FBI man tailing me."

Shows sat in silent surprise.

Von Braun said, "Look behind us."

Shows did as he was told.

"That third car behind us has been tailing us since we left the studio. It's FBI men—they're always around. My telephone is bugged, and the FBI reads more of my mail than I do! I hope some day they'll trust me and leave me alone."[9]

Von Braun may have been wrong about who was tailing him or Shows's memory may have been wrong about von Braun indicting the FBI, but von

Braun was under regular surveillance. While the FBI compiled a large file on von Braun, it focused on matters relating to security clearances and von Braun's personal security; it contained no mention of surveillance on von Braun, tapping his phone, or reading his mail. Army Intelligence files, on the other hand, have numerous references to surveillance on von Braun and checks of his mail. It appears that von Braun's employer, the Army, did not trust him. Still, von Braun had ways of finding privacy.

Charles Shows was as impressed by Wernher von Braun's physical energy as he was by the man's technical genius. He wrote:

> On several occasions he worked nonstop for 12 hours at the studio, then took a taxi from Hollywood to Long Beach, about 50 miles away. There he rented a speedboat and piloted it 22 miles across the water to Catalina Island. Once there, von Braun skin-dived all night, then piloted his speedboat back to Long Beach at dawn. He rode a taxi to his hotel for a quick shower, then reported back to Disney Studio for another long day's demanding work on space travel.[8]

One can imagine von Braun's sense of satisfaction as he got into the speedboat and disappeared into the darkness of the Pacific, leaving the surveillance team standing on the dock at Long Beach. One can sense von Braun's delight at going somewhere no one could find him or watch him, deep into the kelp forests in the clear waters surrounding Santa Catalina Island.

As the wealth of ideas turned into program material, the original concept for one show, "Rockets and Space" turned into two shows, called "Man in Space" and "Man and the Moon." Then the Disney team added a third film, "Mars and Beyond."[6]

"Man in Space" and "Man and the Moon." Let's be honest about the terminology. When Disney and von Braun said "Man," they meant *Man*. A woman's job in the early 1950s, if she worked, was to make coffee, serve cakes, and keep the classical music playing to make a comfortable environment for the men who were doing the real work. A woman's place was in the home, where Lillian Disney and Maria von Braun spent their time taking care of the kids. A woman's place in the years to come was to sit at home in stoic bravery as their husbands embarked on the greatest adventure of the century.

Within months of its first appearance on ABC, *Disneyland* became the most popular hour on television. So, on March 9, 1955, when the "Man in Space" segment appeared, tens of millions of Americans were hunkered down in front of their twelve-inch, black-and-white consoles for their weekly ration of movie-quality entertainment.

Walt Disney appears on the screen to introduce the program. The objective of this evening's show, he says, is "to combine the tools of our trade with the knowledge of scientists to give a factual picture of the latest plans for man's newest adventure." As the show moves on, Disney introduces Ward Kimball, the producer of the show, who in turn introduces his technical wizards, Willy Ley, Heinz Haber, and Wernher von Braun. The remainder of the program takes on a clear German accent as they explain, with the occasional aid of animated cartoons, how man will conquer space.[7,10]

Willy Ley leads off with a history of rocketry. Heinz Haber then discusses the physical problems man has to deal with in space, such as weightlessness. When his turn comes, Wernher von Braun seizes command.[7] He speaks in an assured, though high-pitched voice. His English is precise, and his soft German accent surprisingly melodic to American ears. He describes his plans for sending man into space. "If we were to start today on an organized, well supported space program, . . . I believe a practical passenger rocket could be built and tested within ten years." Then, letting everybody know that he is the world's leading authority on rockets and space exploration, von Braun gets to the specifics: "Now here is my design for a four-stage orbital rocket ship."

The Disney-von Braun rocket is similar in design to the *Collier's* rocket, although it has a few significant differences. The *Collier's* articles proposed a three-stage rocket, but to avoid a potential problem of copyright infringement, von Braun and Disney add an unnecessary fourth stage. Also, the main booster stage has broad stabilizing fins, and the fourth stage, which will return its passengers to earth, has a broad delta wing.

"First," von Braun says, "we would design and build the fourth stage [the orbiting vehicle that would carry passengers] and then tow it into the air to test it as a glider. . . . This is the section that must ultimately return the men to the earth safely."

Disney's animators then take over for the conclusion of "Man in Space." The narrator describes the scene as a "small atoll of coral islands in the Pacific where man is dedicated to just one cause—the conquest of space." Von Braun's four-stage rocket is seen against the predawn sky illuminated by floodlights, prepared for launch. Sirens sound a warning. Technicians monitor their consoles prepared for the imminent firing, (a scene that for a time in the 1960s and 1970s became commonplace.) The narrator continues, "Now man will bet his life against the unknown dangers of space travel."[10]

It was reported that nearly 100 million people watched "Man in Space" when it was first broadcast on March 9, 1955,[6] though a more conservative estimate was 42 million. Those who missed it or wanted to see it again got a second chance when it was rerun on June 15, 1955.[7]

An often-told story illustrates the impact of the first showing of "Man in Space." According to the story, the morning after "Man in Space" was seen on national television, President Eisenhower called Walt Disney to compliment him on the show, and to ask for a print of it "to show all those stuffy generals in the Pentagon" what kind of planning they should be doing regarding outer space.[11] Whether or not Eisenhower actually made this request, he did show some interest in space exploration, although clearly defined and limited. On July 29, 1955, he announced that the United States would launch an artificial earth satellite during the International Geophysical Year (IGY), which was to begin in July 1957 (see Chapter 8).

Not long after Eisenhower announced his intention to launch a satellite, Ward Kimball wrote to Wernher von Braun stating that Disney intended to publicize its second *Disneyland* television show on space by claiming that the first show of the series prompted Eisenhower's decision. Von Braun was aghast. "For God's sake don't put it that this show triggered the presidential announcement."[10] Von Braun could imagine Eisenhower's unhappy reaction being passed down the chain of command until it landed heavily on his shoulders.

News of the "Man in Space" program traveled as far as the Soviet Union, where it was not televised. Professor Leonid I. Sedov, who emerged in later years as chairman of the commission for space flight programs in the USSR, wrote to Frederick C. Durant III, president of the International Astronautical Federation, asking for his help in obtaining a print of the film. He wrote, "If the Disney Studios supplies us with one copy of this film on whatever terms it may put, it will make considerably for the cause of promoting our contact."[10] It is unlikely that Sedov got his wish since the Disney Company was always intensely protective of its creations and since Walt Disney himself was fiercely anti-Communist.

The second *Disneyland* program on space, "Man and the Moon," reached television screens on December 28, 1955. Von Braun made an appearance on the show to present his plan for reaching the moon in two steps: first, by building an orbiting space station that would be a staging area for the lunar trip, and second, by the trip from the space station to the moon and back.

The crew in space (the term astronaut has not yet been coined) work in "bottle suits" of von Braun's design: tiny, one-man spacecraft with seven control arms, to assemble the station. The Disney–von Braun space station is a derivative of the one that appeared in *Collier's* three years earlier; it has the same shape and size and it orbits at the same altitude, 1,075 miles.[12]

"Our space station," von Braun instructed his millions of viewers, "will have the shape of a wheel measuring 250 feet across. The outside rim will contain living and working quarters for a crew of 50 men. Just below the radio and radar antenna is an atomic reactor. Its heat will be used to drive a turbo generator which supplies the station with electricity."[10]

After the space station has been built and served its function as a staging point, attention shifts to the moon. The flight to the moon is simulated by live actors using props based on von Braun's designs. During the voyage, a meteor dramatically strikes the moon ship, and a crew member enters a bottle suit to make emergency repairs.[10] The Disney–von Braun moon ship circles around the moon's unseen side but does not land. No one knows if a suitable landing site can be found, if the moon's surface is craggy peaks and boulder fields, or if the moon is blanketed with fields of light cosmic dust that will, like quicksand, swallow the spaceship. The spaceship is turned by the moon's gravity and makes its return trip to earth.[7]

The third of the *Disneyland* space exploration series titled "Mars and Beyond" was broadcast on December 4, 1957. Its technical advisors were Wernher von Braun, E. C. Slipher, an astronomer from the Lowell Observatory in Arizona, and Ernst Stuhlinger, a member of von Braun's Huntsville-Army team. With Walt Disney's approval and encouragement, von Braun envisioned a journey to the red planet in spaceships driven by a radically new, but enormously efficient atomic-powered engine. Stuhlinger was an authority on the yet to be built engine.[10]

"Mars and Beyond," in standard Disney style, features animation and the use of model spacecraft. With explanations by von Braun and Stuhlinger, the program shows six umbrella-shaped spaceships being assembled in earth orbit, some equipped with landing craft for the exploration of the Martian surface. The round-trip to Mars and back takes thirteen months and six days, a significant improvement over the sixteen months needed by the expedition von Braun proposed in *Collier's* four years earlier.[12,13]

Disneyland opened its doors on Sunday, July 17, 1955, a year and a day after construction began. Disney invited 11,000 people to the opening, and 28,154 showed up—the overflow the result of forged passes. Many attractions were not yet ready. The ABC television network brought Disneyland's first day to those without tickets in a ninety-minute broadcast. Television personalities Art Linkletter, Bob Cummings, and Ronald Reagan guided viewers through an unrehearsed tour of the park that some have characterized as an embarrassing disaster, but that others saw as a window to something exciting and prophetic.[14]

Wernher von Braun and Ward Kimball were the main creative contributors to what existed of Tomorrowland.[15] Its centerpiece was a rocket that stood seventy-six feet high on three shock-absorbing legs. It had the smooth sweeping lines of von Braun's V-2. It was gleaming white with red trim and had, in slanting script near the top, TWA. Trans World Airways paid part of the cost of the attraction for the privilege of advertising itself as the most forward-looking airline on earth. The "Rocket to the Moon" attraction occupied a building behind TWA's rocket. It was all rounded shapes and sweeping curves: the 1950s version of the architecture of tomorrow.

The passengers on the "Rocket to the Moon" enter a circular theater that appears to be the interior of a large rocket ship. Three tiers of seats rise in concentric rings facing the center of the passenger compartment. On the floor at the center is a projection screen showing where the space ship has been. Above seats, on the ceiling, a second projection screen shows where the space ship is going.

As the rocket engines fire, flame fills the screen on the floor projection screen. The air cushions of the theater seats deflate, and the passengers sink into them as if pushed there by the force of the rocket's acceleration. Then the floor screen shows an aerial view of Tomorrowland, Anaheim, and the rest of California, the earth shrinking as time and distance increase. On the projection screen above, the bright blue California sky dims to the blackness of space, punctuated only by the occasional star. The moon appears and grows larger as the spaceship approaches it. While journeying to the moon, the passengers are instructed about the moon and the planets by a voice on the public address system.

The rocket ship will not land on the moon for the same reason the moon ship in "Man and the Moon" did not land; it is not known if a landing can be made safely. The spaceship circles around the unseen side of the moon and returns to earth. The trip back is fast, and the launching-landing pad appears on the floor projection screen.

A recorded announcement thanks the passengers for taking the trip to the earth's nearest neighbor, and reminds them not to forget to take their belongings when they exit the space ship.[16,17,18]

Tomorrowland with its ersatz spaceship and "Rocket to the Moon" attraction was the nation's only space port for about five years, until the Mercury astronauts blasted off from Cape Canaveral on Redstone rockets. It was also a great promotion for those who wanted to explore space or go to the moon. Thousands took the "Rocket to the Moon" daily; before long the passengers numbered in the millions. With their interest made tangible by the combined genius of Wernher von Braun and Walt Disney, people were ready for the real thing.

While von Braun consulted for Disney and made many creative contributions to the Tomorrowland projects, he was still a civilian employee of the United States Army and not yet a citizen. He was responsible for directing development of the Redstone and other Army rockets. The Army reminded him of his position by occasionally tugging on his leash. He continued to be under surveillance. While von Braun may have viewed this as a matter of trust,[9] the Army also had an underlying concern that he might be kidnapped and reappear on the wrong side of the Iron Curtain. In early 1954, just before he became involved with Disneyland, von Braun requested permission to attend the Fifth Congress of the International Astronautical Federation at Innsbruck, Austria, as the official representative of the Army Ordnance Corps. His request, including a commitment to pay his expenses was approved at a lower level, but Brig. Gen. Holger M. Toftoy, commanding officer of the Redstone Arsenal, turned it down because of the risk to von Braun's personal security during the trip, which included stops in England and Germany as well as Austria.[19]

Security clearances were also an issue. As an alien working for the Army on weapons-related projects, Wernher von Braun had been the subject of security investigations since 1947, when he worked for the Army under Project Paperclip. On October 15, 1954, he applied to the Army for renewal of his security clearance at the level of "secret"[20]; six weeks later he was in a chair with sensors strapped to his body for a polygraph test. The report of the test, dated December 3, 1954, stated that it measured no reactions indicative of deception. However, it also noted that von Braun's father-in-law, Alexander von Quistorp, was "still in a prison camp in the hands of the Communists," a suggestion that von Braun might be susceptible to blackmail.[21]

On April 14, 1955, soon after the first showing of the "Man in Space" episode on the *Disneyland* television show, Wernher von Braun became a citizen of the United States. Forty German rocket experts including von Braun, their wives and children, a total of 103 people, took the oath of citizenship at the Huntsville High School auditorium.[22] Twelve hundred people were present for the ceremony, and the mayor of Huntsville declared it New Citizens Day. To Huntsville's new citizens he said, "I am glad you have chosen us. I know of no group we have enjoyed joining our community more." The chairman of the Madison County Board of Commissioners said, "this occasion adds new vitality and strength to our community."[23]

As Huntsville celebrated, Wernher von Braun was reflective. "This is the happiest and most significant day of my life," he said. "I must say we all became citizens in our hearts long ago. I have never regretted the decision to come to this country. As time goes by, I can see even more clearly that it was a moral decision we made that day at Peenemuende."[22]

8

The First Race for Space

> President Eisenhower announced that plans had been approved for "the launching of small unmanned Earth-circling satellites" as part of America's contribution to the International Geophysical Year.
>   Almost immediately, the program went wrong.
>
> Wernher von Braun[1]

Until the middle of 1954, the concept of an artificial earth satellite was just that, a theoretical concept that existed only on paper, on the fiction pages of magazines, or the television screen. In late June 1954, Wernher von Braun became involved in the actual planning of a project to send a small package of scientific instruments into orbit around the earth. The project would be sponsored and paid for by the United States government.

America's first tentative steps toward space were taken by the United States Navy. As early as October 1945, the month that Wernher von Braun arrived at Fort Bliss, the Navy established a Committee for Evaluating the Feasibility of Space Rocketry (CEFSR) within its bureau of aeronautics. The Committee recommended that the Navy begin development of a satellite to carry scientific instruments into orbit around earth, and then awarded a contract to the Guggenheim Aeronautical Laboratory at the

California Institute of Technology to evaluate the basics of carrier vehicle performance, satellite size, and orbital height.

At about the same time, the Office of Naval Research and the Naval Research Laboratory were studying the feasibility of a high-altitude rocket research program. Their study led to the Viking project and the awarding of contracts for the production of twelve rockets.

By early 1946, the Guggenheim Aeronautical Laboratory reported back to the Navy with an analysis that demonstrated that orbiting a satellite was beyond its financial—if not its technical—capabilities. The Navy's CEFSR approached as a potential partner the Army Air Corps on March 7, 1946. The Army Air Corps was noncommittal. Its representatives did not tell the Navy that it had already contracted with the Rand Project (a part of Douglas Aircraft Company that later became the Rand Corporation) to study the feasibility of artificial earth satellites, and the Army Air Corps had no intention of collaborating with the Navy.

On May 12, 1946, Rand outlined to the Air Materiel Command a proposal for a 500-pound satellite to be launched by a rocket based on the V-2. Rand also suggested that the satellite could carry instrumentation for meteorological research, communications, and reconnaissance. The cost was a prohibitive $150 million. Rand then put a finer point on its pencils, pulled out its slide rules, and one year later came back with a proposal for a smaller satellite that could be put into orbit for $82 million.

In late 1947, the Department of Defense, through its Committee on Guided Missiles of the Joint Research and Development Board, reviewed the feasibility of the proposals of the Navy and the Air Force, which was created from the Army Air Corps in July 1947. In all the discussions, the focus was on the concept of putting an artificial satellite into orbit around the earth. Little thought was given to what the satellite would do. There had to be a purpose for building an artificial satellite to justify the enormous cost. Vice Chief of Staff Hoyt S. Vandenberg placed artificial satellites in their proper perspective when he said that "satellites should be developed at the proper time." But the United States did not have a rocket—or plans for one—powerful enough to do the job.

Apparently neither the Navy nor the newly created Air Force considered a collaboration with the Army, which had in its employ the world's leading authorities on designing, building, and firing rockets capable of entering space. To avoid being left out of its own area of expertise, on September 15, 1948, Army Ordnance submitted a proposal to the Department of Defense's Committee on Guided Missiles for a continuing study of the feasibility of developing a satellite based on the work of Wernher von Braun's team and their V-2.[2,3]

At the end of 1948, Secretary of Defense James V. Forrestal wrote the following in his report on the activities of his department:

The Earth Satellite Vehicle Program, which was being carried out independently by each military service, was assigned to the Committee on Guided Missiles for coordination. To provide an integrated program with resultant elimination of duplication, the committee recommended that current efforts in the field be limited to studies and component designs: well defined areas of such research have been allocated to each of the three military departments.[2]

Forrestal's commentary could have been viewed as cautious yet promising. But nothing more was heard of satellite development in the U.S. military services for over five years.

While the U.S. military services let the Earth Satellite Vehicle Program slide into oblivion, Wernher von Braun refused to give up. While he envisioned peaceful space exploration in his series of visionary articles in *Collier's*, he wrote an article for *Ordnance* that more closely reflected the political climate of the Cold War. In the article, "Space Superiority," Wernher von Braun played the hawk in contrast to the role of dove of the *Collier's* articles.

The atom bomb may have been the "ultimate weapon" heretofore, but this will not be the case much longer.

It is now the time . . . when we must relegate strategic bombing to a secondary position and seek a new "ultimate weapon" which shall preferably not only return to us that deciding "edge" we once had over Red aggression but likewise kinder to the taxpayer.

We need fear no fundamental, immovable roadblocks along the path of further development of the large liquid [-fueled] rocket, as the principal difficulties have already been conquered. At some distance along this path—a very attainable distance, by the way—stands the multistage, orbital rocket ship bearing a crew out beyond the stratosphere.

The first nation to launch such a rocket ship will possess, in my opinion, what may well be the long-sought "ultimate weapon."[4]

Von Braun's ultimate weapon culminated in the construction of a permanently manned earth satellite, a space station.

The first and obvious military application of such a station is that of reconnaissance and observation.

The station would also operate improved radarscope equipment far superior to the present aircraft-borne devices and thus be able to penetrate reliably the thickest cloud overcast.

An orbital reconnaissance station can pull up an iron curtain![4]

Reconnaissance was relatively benign compared with von Braun's next proposal.

> When it comes down to cases, the outstation is also a launching plat-form for orbital missiles against which there cannot well be counter-measures. If we fire an atom-tipped, winged rocket backward from the outstation so that its thrust diminishes its orbital velocity relative to the station by 1,070 miles an hour, it will succumb to gravity and approach the earth on an elliptical path.[4]

The apocalyptic denoument of such a drama was obvious, though von Braun described the descent to earth of his nuclear missile in expansive detail.

Von Braun conceded that building his proposed military space station would be a Herculean task. Nevertheless, he claimed that it could be op-erational within ten years at a cost of less than $4 billion. While von Braun had fantastic vision, he was also a realist. He concluded his article with a pitch for a small, earth-orbiting satellite as a means of solving practical problems in the development of the military space station.[4]

The practical planning for the United States's first artificial satellite began in June 1954, when Wernher von Braun received a phone call from his friend Frederick C. Durant III, former president of the American Rocket Society and in 1954 the current president of the International Astronautical Federation. Durant had just had a conversation with a man from the Office of Naval Research who wanted to start building an artificial earth satellite, and Durant wanted to know if von Braun would talk to the man.

Durant organized a meeting in Washington, D.C., to take place on June 25, of a small group of experts who subscribed to the idea that the time of the artificial satellite had come. Those in attendance were Frederick C. Durant III; Commander George W. Hoover of the Navy, the man whose interest initiated the meeting; Alexander Satin, an engineer from the Office of Naval Research; S. Fred Singer, a physicist from the University of Mar-yland; Fred L. Whipple, an astronomer from Harvard University; David Young, of the Aerojet-General Corporation; and Wernher von Braun. After a brief introduction by Durant, George Hoover took the floor and gave the meeting its direction. He called for placing a small satellite into orbit by using a combination of existing rockets.[5]

The participants in the meeting, all believers in the need for an artificial satellite, quickly sketched out a design for the rocket stages and the satellite. Since Hoover called for the use of off-the-shelf components, much of the launch vehicle design was predetermined. The largest booster rocket avail-able at the time was the Army's Redstone, designed by von Braun's team. For simplicity, upper stages would be built by clustering small, solid-fueled rockets. Depending on the rockets used for the upper stages—those proposed were also from the Army's arsenal—the vehicle could launch a satellite weighing at least five, or as much as fifteen to twenty pounds.

Commander Hoover carried the ad hoc group's design ideas back to the chief of Naval Research who approved the concept and authorized further discussion with the Army's rocket development group at Redstone Arsenal. Project Orbiter was born, and at last it seemed as if interservice cooperation would get a satellite off the ground.[6]

The Navy's representatives came down to Huntsville on August 3 to meet with Wernher von Braun and Holger Toftoy to set up the Army-Navy joint venture. Von Braun and, more important, Toftoy, were sold on the idea. Toftoy got the approval of his superiors under the condition that the artificial satellite project did not interfere with other Army rocket development programs.

Von Braun took on the responsibility of formalizing the satellite proposal in writing. The title of his secret report, dated September 15, 1954, was "A Minimum Satellite Vehicle Based on Components Available from Developments of the Army Ordnance Corps." He estimated the cost to the Army at $100,000. The proposal included not only a technical description but also evidence of von Braun's growing global political savvy. He wrote that

> The establishment of a man-made satellite, no matter how humble, e.g., five pounds, would be a scientific achievement of tremendous impact. Since it is a project that could be realized within a few years with rockets and guided missile experience available now, it is only logical to assume that other countries could do the same. *It would be a blow to the U.S. prestige if we did not do it first* [von Braun's emphasis].[7]

In order to present a unified program to the Department of Defense and to maximize the chances of approval, the Army and Navy invited the Air Force to join Project Orbiter in January 1955.[6] They also submitted a proposal for the satellite plan to the assistant secretary of defense for Research and Development on January 20, as a first step in gaining the approval of the secretary of defense and the president.[7]

As groups within the United States Army and Navy planned to launch a satellite, scientists in other countries made similar plans. Several international scientific societies began organizing for what became known as the International Geophysical Year (IGY). The year was, in fact, eighteen months long, stretching from July 1, 1957 to December 31, 1958. Ultimately, sixty-seven groups joined the IGY enterprise to broadly study the nature of the earth and the sun.[8] When a special committee of the IGY met in Rome on October 4, 1954, it recommended that thought be given to launching an artificial satellite during the IGY. The IGY went so far as to create a logo, approved in 1955, which showed a small satellite in orbit around the globe.[9]

On March 14, 1955, the United States National Committee for the IGY approved the satellite concept. The Department of Defense took note of all the pronouncements of the civilian scientific societies, and in May 1955, it created the ad hoc Advisory Group on Special Capabilities to study the feasibility of an artificial earth satellite.[6,10] On July 29, 1955, the National Academy of Sciences and the National Science Foundation announced that the United States intended to orbit an artificial satellite during the IGY, and the same day President Dwight Eisenhower climbed on board when he announced that an artificial satellite would be part of the United States IGY program.[11] The following day, July 30, the Soviet Academy of Sciences also announced that it intended to launch an artificial satellite.[12] It is noteworthy that the various scientific organizations made recommendations, but only President Eisenhower had the resources and authority to authorize the building and orbiting of a satellite.

By that time, the military services' charade of cooperation had fallen apart, and the Department of Defense's Advisory Group on Special Capabilities had three possible systems from which to choose. The Army submitted the concept of Project Orbiter with von Braun's design of the Redstone as the first stage and the upper stages composed of clusters of solid-fueled rockets. Von Braun assured the advisory group that the Army could orbit a fifteen-pound satellite by mid-1956. The Navy presented a proposal that originated in the Naval Research Laboratories that required a new launch vehicle they called Vanguard. The first stage of the Vanguard was an upgraded Viking sounding rocket, one carrying instruments to analyze the upper atmosphere; the second stage was a modified Aerobee-Hi sounding rocket, and the third stage was a completely new design. The Navy claimed it could orbit a forty-pound satellite in the same time frame proposed by the Army's Project Orbiter. Not to be outdone, the Air Force submitted a proposal to launch a very large satellite with its Atlas ICBM as a booster. The Atlas, however, encountered development difficulties and was not considered a serious candidate.[10]

On September 9, 1955, the Advisory Group on Special Capabilities made its decision, and while members of the committee never fully explained their choice, there was no shortage of speculation about their decision. President Eisenhower insisted that the satellite project be scientific, not military. Who knew what reaction foreign nations might have on discovering a ballistic missile flying over their territories, even if it was harmlessly carrying a satellite into orbit. This concern immediately put the Army and Air Force programs at a near fatal disadvantage. There was concern that the artificial satellite be seen as an American project, not as a project funded by the United States government and effected by a team of Germans.[13] Also, there was talk that the Army's Project Orbiter presented an inelegant solution, since it was based on technically sound, but aesthetically deficient military hardware.[7] The Department of Defense's advisory group voted seven to two to launch a satellite with the Navy's Vanguard vehicle, and the Department of Defense and the Administration approved the recommendation.[6]

While the Army was not to participate in the Vanguard satellite program, the Department of Defense did not preclude its participation—in a yet to be defined way—in its space initiative. It kept the Army involved, at least from a public relations perspective, by enlisting its participation in an informational and educational film, *The Challenge of Outer Space*, which was released on February 29, 1956.[14] The sixty-one-minute-long film featured Wernher von Braun, who, according to the Department of Defense's description of the film, defined "the characteristics of present day guided missile models, how they function, the problems involved in achieving operational effectiveness, and the challenge presented by outer space in future development."[15]

Although the Army lost its bid to orbit the first satellite for the United States, it won a consolation prize from the Department of Defense. In July 1955, von Braun's group at the Redstone Arsenal proposed to develop an intermediate-range (1,500 miles) ballistic missile (IRBM) that could deliver a 2,500 pound payload. It would be named the Jupiter. While the Air Force had for some years been funding the development of the Atlas intercontinental ballistic missile (ICBM), its successful deployment was not a sure thing, and development of a shorter-range missile by a group that had proved its capabilities seemed prudent. The Department of Defense approved a program that resulted in a missile system operated from land by the Army and from the deck of ships by the Navy. To cover its bets, the Department of Defense also approved the Air Force's program to develop the Thor IRBM.[16]

With its new assignment, the Redstone Arsenal got a new organization and von Braun's group acquired a new, more prestigious title. On February 1, 1956, the Army establishment at Redstone became the U.S. Army Ballistic Missile Agency (ABMA) under the command of Brig. Gen. John B. Medaris. Wernher von Braun became director, Development Operations Division, ABMA, and reported to Medaris. Brig. Gen. Holger Toftoy, who brought von Braun to the United States and was his mentor over the years, kept command of the remainder of Redstone, which operated under the name U.S. Army Rocket and Guided Missile Agency.[16,17]

Although the Navy had won the first lap, Wernher von Braun and his team at the Redstone Arsenal were far from out of the race for space. Von Braun and Redstone scored a major coup in July 1956, when, after years of negotiation, von Braun managed to bring Hermann Oberth to Huntsville to work on special projects. As one of the first theoreticians of space travel, Oberth was a senior statesman of space exploration, and he was, in his irascible way, still pouring out creative ideas.[18]

The Jupiter that von Braun and his team finally built was an all new missile with a range of nearly 2,000 miles. It was sixty feet one inch long and eight feet nine inches in diameter. It weighed 110,000 pounds and was powered by an engine that produced 150,000 pounds of thrust.[19] While the Jupiter was an entirely new missile, von Braun and his group planned to use the reliable old Redstone as a test vehicle for components. Critical to the design of the Jupiter was a nose cone that traveled accurately to its target and did not burn up as it reentered the earth's atmosphere. The Army's approach was to use as the outer surface of the nose cone an ablative material that dissipated heat as it burned away on reentry. To test the concept, von Braun arranged to build a dozen test vehicles called the Jupiter C. The Jupiter C bore little resemblance to the Jupiter IRBM. It was composed of a much smaller, upgraded Redstone booster stage with upper stages built from clusters of identical, small solid-fueled rockets. The second stage had a cluster of eleven rockets, the third stage was a cluster of three rockets that fit within the ring formed by the eleven rockets of the second stage. Atop it all was the fourth stage, the test nose cone, a scaled-down version of the nose cone intended for the Jupiter IRBM. Although it had a different name, in concept the Jupiter C was the same vehicle that von Braun had proposed for Project Orbiter.[20] This similarity did not escape notice at the Pentagon. Before the first scheduled launch of the Jupiter C, Brigadier General Medaris was ordered to personally inspect the fourth stage to insure that it contained no rocket fuel. The Pentagon could not allow von Braun and his team to blast a satellite into space and apologize after the act.[21]

The first Jupiter C blasted off from Cape Canaveral on September 20, 1956, carrying its payload to an altitude of 682 miles and covering a distance of 3,400 miles. Although the nose cone was not recovered, the altitude and distance record stood for two years.[20]

The Jupiter C program accomplished its stated goals after firing only three of the twelve approved rockets. The remaining rockets were in varying stages of assembly, and Brigadier General Medaris ordered the hardware placed into protected storage. Two of the vehicles were kept in near flight ready condition, in case they were needed on short notice.[22]

Not long after the successful first flight of the Jupiter C, Secretary of Defense Charles E. Wilson acted to end the internecine battles of the military services under his direction for primacy in the development of rockets and, by implication, access to space. The redundant programs simply cost too much. Therefore, on November 26, 1956, Wilson issued his Roles and Missions Directive: it limited Army development activities to rockets with a range of not more than 200 miles. The Air Force and Navy were responsible for missiles with longer ranges, the IRBMs and ICBMs.[23]

The Air Force, which was simultaneously developing its own Thor IRBM had little use for the Army's Jupiter IRBM. Nevertheless, von Braun's ABMA group at Huntsville completed development of the Jupiter and turned it over to the Air Force for deployment. The Air Force deployed two squadrons, each with 30 missiles, one to Italy and the other to eastern Turkey.[19] The Jupiters were out of sight and out of mind to the Air Force, but the Soviet Union could not help but notice that the missiles could reach as far west as its European satellites, as far east as Tashkent, and almost as far north as the Barents Sea. Stalingrad, Moscow, and—with a stretch—Leningrad were within their range.

According to the development plan, when the Jupiter missiles were turned over to the Air Force, the Army, its ballistic missile agency at Redstone Arsenal, and Wernher von Braun would be officially out of the space business.

9

The Soviet Chief Designer and Sputnik

> I suspect we will have to pass Russian customs when we finally reach the moon.
>
> Wernher von Braun[1]

Every hero in a grand drama should have a foil, a counterpart who, by contrast, accentuates his qualities of greatness. Wernher von Braun had such a foil, a man who, like himself, dreamed of building spaceships powered by rockets, who designed ballistic missiles for his country, and whose work was assisted by a team of German rocket scientists who learned their trade in service to the Third Reich. The man was a shadowy figure, unknown both in his country and outside of it. When von Braun's foil was first spoken of, after the launching of the first artificial satellite to orbit the earth, he was known only as the "chief designer" of the Soviet Union.

In the Communist culture of the Soviet Union, the individual had value only to the extent that he or she contributed to the whole of society. The genius of an individual, if allowed to be expressed, became submerged in the whole as an achievement of the Soviet people. At least that was the theory. The contributions of the chief designer who inspired and led the Soviet space program were arguably greater than Wernher von Braun's contributions to the United States space program. The anonymous chief designer of the Soviet space program was Sergei Pavlovich Korolyov (also Korolev).

Long before Wernher von Braun and Sergei Korolyov made space travel a reality, even before Robert Goddard and Hermann Oberth created liquid-fueled rockets and charted the way to the stars, a Polish-born Russian named Konstantin Eduardovich Tsiolkovsky (1857–1935) laid the foundation for space travel.

Tsiolkovsky was a high school mathematics and physics teacher who, in 1883, began to puzzle about how one might travel through space. Twenty years later, in 1903, he finally described his studies in a paper, "Investigation of Cosmic Space by Reactive Machines," which he published in the Russian journal *Scientific Survey*. A reactive machine exploited Newton's third law of motion, which said that for every action there is an equal and opposite reaction. In Tsiolkovsky's mind, a reactive machine was a rocket. His later work proposed the modern concepts of the multistage rocket and a rocket motor fueled by liquid hydrogen and liquid oxygen. Unfortunately, because of his lack of financial resources, Tsiolkovsky was not able to build a rocket. His theoretical studies and proposals were all published in Russian. They were not translated or available outside his native land.[2,3] They were not read, except by other, younger Russians like Sergei Korolyov.

Photographs of Korolyov show him to have a heavy physique, more muscular than fat, with agility and strength despite its bulk. He had brushed back dark hair and a rounded face with sharp features. He bore a superficial resemblance to Wernher von Braun, who was only five years his junior. Those who knew him used the same words to describe him that were also applicable to von Braun: ambitious, passionate, brilliant. Like von Braun, there was a part of his past that he would not discuss; Korolyov had his own experience with prison camps.[4]

The story of Sergei Korolyov, like much of Soviet history, is clouded by the machinations of Soviet policy. The Soviet Union created myths, deleted embarrassing truths, and omitted what it considered irrelevant. Truth always took second place to an expedient story. So it was with the biography of Korolyov. Even after the demise of the Soviet Union much of his story is lost and the facts are difficult to verify.

Sergei Pavlovich Korolyov was born in 1907 in Ukraine, where he spent his early years. In Moscow, he learned his trade as an aircraft designer from the Russian master Andrei Tupelov, and earned his living as an aeronautical engineer. By the early 1930s, Korolyov became interested in rockets as a means of propulsion and space travel.[5] He displayed his knowledge and expertise in 1932 when he wrote a pamphlet, *Rocket Motors*, published by the Soviet military establishment. He was also an active member of the Group for the Study of Rocket Motors, commonly known by its Russian initials, GIRD, which was the Soviet Union's equivalent of Germany's VfR and *Raketenflugplatz* groups.[6] Korolyov was reportedly a dynamic leader of GIRD. He convinced Soviet Armaments Minister Mikhail Tukhachevsky of the potential of rockets as weapons, and obtained the powerful man's support for GIRD in the form of a rocket-launching base near Moscow. There, in 1933, GIRD triumphantly fired the Soviet Union's first liquid-fueled rocket, a device that weighed thirty-eight pounds and reached an altitude of 1,300 feet.[7]

On October 31, 1933, the Soviet government merged GIRD with the Leningrad Laboratory for Gas Dynamics to form the Institute for Research into Jet Propulsion. In effect, GIRD became part of a state institution and a well-funded part of the military establishment. On November 9, 1933, Sergei Korolyov became deputy head of the institute with the title of divisional engineer; his civilian job title was equivalent to the rank of general in the Soviet Army. With the Soviet military supporting him, Korolyov began two projects, the first, a rocket with an intended range of thirty miles and the second, a glider equipped with a liquid-fueled jet motor.[8]

In 1934, Korolyov visited Konstantin Tsiolkovsky, the patriarch of Soviet rocket science. Before his death in 1935, Tsiolkovsky passed on his legacy and dream to younger men like Korolyov who had the government's support and were in a position to make the dream a reality.[9]

Korolyov's good fortune ended on June 10, 1937, when Soviet leader Joseph Stalin, in a paranoid exercise of power had the NKVD (secret police) arrest his armaments minister, Tukhachevsky, and all senior staff members under him. Virtually all of them perished in the purges of 1937 and 1938, except Korolyov. For reasons still not known, Korolyov was sent deep into a Siberian gulag to serve a life term for treason. He spent a year slaving at the Kolyma gold mines of eastern Siberia, in one of the most soul destroying and murderous of the forced labor camps.[8,10] Korolyov left behind his wife and young daughter, who was born in 1935. According to one account, his wife yielded to pressure and denounced him while he was at Kolyma.[11]

Korolyov would have eventually died in the Kolyma gold mines had it not been for his former teacher and boss, Andrei Tupolev. Tupolev, his wife, and his entire staff had the misfortune to be arrested by Stalin's secret police in 1938, but the good fortune to be sent to a *sharashka*, a prison where conditions were relatively good. They could, in fact, were required to continue their work. To round out his team, Tupolev somehow managed to find Korolyov and bring him to the *sharashka* near Moscow.[10,12] Korolyov spent several years there, then, when Germany invaded the Soviet Union, he accompanied Tupolev and his group when they were transferred to Omsk in Siberia. From there, Korolyov was sent to Special Prison Number 4 (location not known), where he worked on small rockets for powering aircraft and on bombardment rockets.[10]

Sometime during 1945, Sergei Korolyov was awarded the Soviet Union's Badge of Honor, "for participation in the development and testing of rocket motors for military aircraft." He was also released from prison, though it is likely that he was kept under close supervision.[13]

Korolyov's new freedom was apparently, in some convoluted way, related to the end of World War II. With the defeat of the Third Reich, Germany became a bazaar filled with military hardware trophies and technical treasures that were greedily scooped up by the victorious powers. By virtue of the division of Germany into zones of occupation, the Soviet Union took possession of Peenemuende and the world's first rocket factory, the Mittelwerk. To the Soviet's dismay, Peenemuende had been stripped of anything of value by von Braun and his group when they evacuated the rocket center during the previous winter. The United States Army, when it briefly occupied the area around Nordhausen, looted the Mittelwerk of about 400 railroad train cars full of rocket parts—enough to build 100 V-2s.[14] The British managed to locate a few intact V-2s, and with the aid of German personnel, intended to test fire them from an artillery range on the North Sea, near Cuxhaven, Germany, under the code name Operation Backfire. The British generously invited their allies to send observers. The Soviet Union sent a team that included Sergei Korolyov.

The British attempted their first V-2 firing from Cuxhaven on October 2, 1945, but because of ignition problems, the missile never left the ground. The following day, October 3, they tried again with stunning success: The missile roared through a four-minute fifty sec. flight and fell to earth one half mile to the left and one mile short of its target.[15] On this same day, the designer of the rocket, Wernher von Braun arrived at his new American home, Fort Bliss, Texas.[16]

On October 4 a second V-2 rose from the launch pad at Cuxhaven. Its flight appeared normal, but its engine burned for only thirty-five seconds, and it fell to earth after covering only fifteen miles.[15]

Sergei Korolyov arrived to observe the third and final flight of a V-2 on October 15. Two high-ranking Soviet officers were allowed into the firing range, but Korolyov was excluded by the British.[15] The Soviets had not submitted his name among those of the official observers, possibly so that Soviet officials could keep him under supervision. Korolyov had to be satisfied with watching the flight from the wrong side of the barbed wire fence.[10] When the V-2 rose into the sky that day, Korolyov may have seen it briefly as it soared above the trees and disappeared into the low clouds. The flight was a qualified success, as the missile fell to earth about a dozen miles short of its target.[15]

Seeing the strategic value of rockets such as the V-2, the Soviets set up operations in Germany to exploit the developments of von Braun and his team and of the Nazis' manufacturing system. They defined two goals: the construction of a complete set of V-2 production drawings, and the resumption of production of V-2 missiles and parts.[17] To accomplish the first goal, they established the *Institut Rabe* (an abbreviation of *Raketenbetrieb*, or Rocket Enterprise) at Bleicherode. For the second, they reopened the old Mittelwerk subterranean rocket factory under the new name of Zentral-werke (Central Works).[18]

The Soviet Union had no difficulty recruiting German experts to assist them in their activities in Germany. The United States Army and Wernher von Braun left behind thousands of skilled and creative men who were unemployed and willing to take the high wages offered by the Soviets. Some of the men had been offered jobs under Operation Overcast but declined them, preferring to stay with friends and family in their homeland, despite its destruction.[19] The most creative and valuable of these men was Helmut Groettrup, a senior engineer on von Braun's team at Peenemuende, who, along with von Braun, was arrested by the Gestapo for allegedly putting their plans for space flight ahead of weapons development for the Third Reich.[20]

It appears that at the higher levels, both Soviet and German personnel had responsibilities that included both rocket design at the *Institut Rabe* and rocket production at the Zentralwerke. In the spring of 1946, Groet-trup was placed in charge of the Zentralwerke and its thousands of German personnel. By summer, he was asked by his Soviet employers to propose technical improvements for the V-2, which he and his colleagues did by mid-September.[19] Korolyov was also assigned to the Zentralwerke to co-ordinate repairs of the subterranean factory and later to ship V-2 equipment back to the Soviet Union.[20] He is also reported to have worked at the *Institut Rabe* on the improved V-2, which would be known as either the K1 or R-1. The missile that emerged from the German design proposals and the Soviet designers was a stretched V-2. The weapon was lengthened by nine feet to accommodate larger fuel tanks, and the engine's perform-ance was improved to increase thrust from twenty-five to thirty-two tons. It's range increased from approximately 200 to 400 miles.[21]

Korolyov and Groettrup reportedly got along well and treated each other with professional respect.[22] The relationship between the Soviets and Ger-mans, while apparently on the best of professional terms, set the pattern for the future. Communication was one way with all the benefits going to the Soviets. The Germans were asked to propose solutions. The Soviets evaluated them, using what they liked. They fostered the notion that the Germans were in charge of their projects and their destiny, both of which would remain in Germany. While Sergei Korolyov and other Soviet rocket scientists took a crash course in rocket design, the Germans never knew if their ideas were used or learned details about Soviet developments.

By 1946 United States Military Intelligence became aware that the Soviet Union was recruiting von Braun's former personnel and putting them to work designing rockets and building new V-2s. It asked Wernher von Braun, through Maj. James Hamill, to evaluate the capabilities available to the Soviets. Hamill's report to Military Intelligence revealed his knowledge of activities in the Soviet Zone of Germany and von Braun's evaluation of his former staff.

> There is no doubt that the bulk of the most capable members of the Peenemuende group are in the United States now. There are, however, many very good former Peenemuende experts working for the Russians, too. In the opinion of Professor von Braun the two most capable of these men . . . are: Dipl. Ing. Helmut Groettrup . . . and Engineer Martin. These two men are, according to the best available information, in charge of the Russian project—new development projects (Groettrup) and the A-4 manufacture in Nordhausen (Martin). [Later reports generally credit Groettrup with overall responsibility for new A-2 (V-2) production at the Zentralwerke.]
> As regards future developments such as A-9, A-10, and A-11 [respectively: a winged V-2 with extended range; a proposed booster for the A-9 capable of giving it intercontinental range; a proposed large booster for the A-9/A-10 combination, which could boost the A-9 into orbit] Groettrup is to be considered a very able and clever leader of a development group.[23]

Hamill quoted von Braun as saying, "I am convinced, without trying to hide the light of our Fort Bliss group under a bushel, that Groettrup will be able to build up gradually a capable group out of former Peenemuende people that can successfully continue these developments for the Russians."[23]

Von Braun further predicted that the Soviets could have the capability of putting the A-10 (the booster capable of giving the winged V-2 intercontinental range) into production by 1949.[24] Von Braun's evaluation of German-Soviet capabilities seems to have been filed away without further response to the Soviet Union's activities or used as a basis for developing United States policy or capabilities.

In the summer of 1946, Sergei Korolyov returned to the Soviet Union to supervise flight tests of captured V-1 buzz bombs at a missile range near the village of Kapustin Yar, about sixty miles east of Stalingrad (now Volgograd) on the lower Volga River.[20,25] When he came back to Germany in the fall, he prepared a list of the most talented German scientists and engineers and turned the list over to his superiors.[20]

By 1946, in the words of Winston Churchill, an Iron Curtain had descended across Europe that separated the former Allies in the war against Nazi Germany and gave formal recognition to their new status as opponents, if not enemies. If another war began, the western Allies held the advantage by virtue of their air bases within striking distance of Moscow and other major cities. The United States alone had ground, air, and naval bases that encircled the USSR. The Americans had demonstrated in the war against Japan that they had atomic weapons and were prepared to use them. The Soviet Union had neither nuclear weapons nor a way to deliver them across the Atlantic. Soviet physicists were developing their atomic bomb, and Sergei Korolyov would build an intercontinental ballistic missile (ICBM) capable of delivering the atomic weapon. This would be done not in Germany, but on Soviet soil.

On October 21, 1946, Soviet General Gaidukov, who was in charge of the Soviet exploitation of German rocket technology, invited Groettrup and his managers to a party to celebrate their collaboration, which had brought about significant improvements in the V-2. A toast by the Soviets was followed inescapably by a toast from the Germans. The cycle was repeated with the best vodka the Soviets had to offer until about 4:00 A.M. on October 22, when Soviet security forces rounded up the hopelessly drunk Germans and their families and put them on board trains destined for Russia.[22] When the Germans arrived, they were broken up into teams and sent to various locations where they worked on problems given to them by their Soviet masters, but the pattern of interaction used in Germany did not change. The Germans did their jobs, but they were never integrated into the Soviet rocket program.

In the fall of 1947, Korolyov directed a series of test launchings of V-2s from the missile range at Kapustin Yar. German personnel were there to show the Soviets how it was done. Before the year was out, they fired about twenty missiles carrying warheads and scientific payloads.[26] In the late 1940s, Korolyov's work with the V-2 and the improved Russian version, the K1 (or R-1) bore fruit. The latter was placed into service as the Red Army Missile.[27]

As Korolyov was having professional success, his personal life was in turmoil. In 1946, he divorced his first wife, Xenia Vincentini, the mother of his only child. The following year he married Nina Kotenkova.[27,28] There are also reports that in 1948 Korolyov was sent back to the *sharashka* (prison). Only the barest details are available, but it could be that under Stalin's repressive regime there was little difference between being in and out of prison.[29]

As the years went by, the German rocket scientists were joined by and then replaced with younger Russian engineers. Finally, between 1951 and 1953, the Germans who had been taken to the Soviet Union against their will were allowed to return to Germany.[30]

At about this same time, Wernher von Braun's deputy at the Huntsville Arsenal, Eberhard Rees, spent a lot of time in Germany recruiting trained personnel to work in the United States Army's missile program. One of the men Rees contacted had recently returned from the Soviet Union and was living in West Germany (the Federal Republic of Germany). He was afraid to talk to Rees because he thought he was under observation by Soviet agents. Another returnee who lived in East Germany (the German Democratic Republic) failed to keep an appointment. Rees assumed that the man had been picked up again by the Soviets. Von Braun had received a letter from a Werner Baum who offered his services to the United States after returning from the Soviet Union. Baum was remembered as a mediocre engineer, and Army Intelligence suspected that he might be working under instructions of the Soviet Union.[31] It was tacitly understood in the United States Army that the Germans who worked for the Soviet Union had little to offer and that hiring them could not justify the security risk.

Years later, when Wernher von Braun evaluated the significance of his former colleagues to the Soviet rocket program, he said, "there is every evidence to believe that their contribution to the Russian space program was almost negligible. They were called upon to write reports about what had happened in the past, but they were squeezed out like lemons, so to speak. In the end, they were sent home without even being informed about what went on in the classified Russian projects."[32]

The Soviet Union was scrambling to build an ICBM that could speed a nuclear warhead to its enemies. Its technologies in both rocket and warhead design were primitive in comparison with those of the United States. Ironically, the Soviets turned these technical deficiencies to their advantage in the first years of what became known as the space race. To begin with, Soviet physicists developed nuclear bombs within a few years of the United States, however, they had not been able to miniaturize them. Consequently, Soviet rocket designers, specifically Sergei Korolyov, had to build rockets far bigger than the Americans had for the same purpose. Bigger rockets later translated into an ability to send heavier payloads into space. Furthermore, Soviet engineers were not able to build more powerful rocket motors that could withstand additional pressure and heat. Korolyov was stuck with a somewhat improved version of the V-2 engine to lift his missiles. He began to design and build missiles that were propelled by cumbersome clusters of these small rocket motors. They were not sleek or beautiful, but they were powerful and they worked.

When Stalin died in 1953, the people of the Soviet Union breathed a collective sigh of relief. Like many others, Korolyov was rehabilitated from his previous position of purge victim: He was invited to join the Communist Party, which he did, and he was elected to the Soviet Academy of Sciences. Not only had he become respectable, he had prestige in the political and scientific subcultures of the Soviet Union.[28]

Former Soviet Premier Nikita S. Khruschev was one of the most reliable witnesses about Korolyov's importance to the development of Soviet rocketry and its space program. In his memoirs he wrote:

[W]hile Stalin was alive he completely monopolized all decisions about our defenses, including—I'd even say *especially*—those involving nuclear weapons and delivery systems.

Not too long after Stalin's death, Korolyov came to a Politbureau [*sic*] meeting to report on his work. I don't want to exaggerate, but I'd say we gawked at what he showed us as if we were a bunch of sheep seeing a new gate for the first time. When he showed us one of his rockets, we thought it looked like nothing but a huge cigar-shaped tube, and we didn't believe it could fly. Korolyov took us on a tour of a launching pad and tried to explain to us how the rocket worked. We were like peasants in a marketplace. We walked around and around the rocket, touching it, tapping it to see if it was sturdy enough—we did everything but lick it to see how it tasted.

We had absolute confidence in Comrade Korolyov. We believed him when he told us that his rocket would not only fly, but that it would travel 7,000 kilometers. When he expounded or defended his ideas, you could see passion burning in his eyes, and his reports were always models of clarity. He had unlimited energy and determination, and he was a brilliant organizer.

Thanks to Comrade Korolyov and his associates, we now had a rocket that could carry a nuclear warhead. His invention also had many peacetime uses. With his Semyorka, he paved the road into outer space.[33]

Korolyov's Semyorka ("Number 7," also known in the USSR as the R-7 and in the West as the SS-6 Sapwood) was an ugly, inefficient brute, poorly suited for its purpose as an ICBM. It was composed of a central rocket with four tapered, strap-on boosters. The central missile was powered by a cluster of four fixed-thrust chambers that were direct descendants of the V-2 rocket engine. The chambers burned kerosene and liquid oxygen, and each produced a little over twenty-five tons of thrust. Each of the four boosters had a cluster of four similar thrust chambers. In addition, the entire assembly had twelve smaller gimbal-mounted rocket engines to control trajectory. All thirty-two of the R-7's main and steering engines fired at liftoff producing over a million pounds of thrust. The R-7 however, was ill suited as a weapon. Its thirst for liquid oxygen necessitated large support facilities and a lengthy fueling time. Its size (structure weight: 61,730 pounds; launch weight: over 650,000 pounds; length: 100 feet; diameter at base: 30 feet) made it hard to transport or hide from the enemy. Its relative simplicity and reliability, however, made it an ideal vehicle for launching large objects into orbit.[34]

The size and power of the R-7 were far beyond the capabilities of rocket-testing ranges in the Soviet Union. In June 1955, to accommodate their new ICBM, the Soviets began construction of a new and secret missile test range near the small village of Tyuratam, northeast of the Aral Sea in Kazakhstan. (For security reasons, after the launch of the first manned satellite, the location of the launch site was given misleadingly as Baikonur, a village hundreds of miles to the northeast. Even today, references to the Baikonur Cosmodrome are common, even though its true location is at Tyuratam.)[35]

Korolyov and his team began test firings of the R-7 in the late spring of 1957, and they went badly. The first missile fired exploded on launching, and several more attempts fared no better. By July, Korolyov was criticized by rival Soviet rocket engineers and bureaucrats. At last, on August 3, 1957, Korolyov and his team accomplished the first successful launch of an R-7. A few weeks later, they fired a second missile into the Pacific Ocean near the Kamchatka peninsula.[36] At last, Korolyov had a heavyweight rocket booster, if not an operational ICBM.

Korolyov explained (in a report published in 1969, after his death) how the concept of the Soviet Union's first artificial earth-orbiting satellite, the Sputnik, came about.

> We followed closely the reports of preparations going on in the United States of America to launch a sputnik called, significantly, *Vanguard*. It seemed to some people at the time that it would be the first satellite in space. So we then reckoned up what we were in a position to do, and we came to the conclusion that we could lift a good 100 kilogrammes (220 lbs) into orbit. We then put the idea to the central committee of the Party, where the reaction was: "It's a very tempting idea. But we shall have to think it over." In the summer of 1957 I was summoned to the Central Committee offices. The "O.K." had been given.[37]

It is likely that Korolyov's request was made and approval given after the first successful flight of an R-7, during the month of August 1957.

Implicit in this account are two significant facts. First, the United States led the "space race" at the time. The leaders of the Soviet Union gave no serious thought to orbiting a satellite until after the United States began its Vanguard program. True, various scientific emissaries spoke publicly about a Soviet space program to come, but it had not been approved or even thought of by those in power. Second, Korolyov tempted the Central Committee and Nikita Khruschev with the idea of outdoing the United States. Propaganda was primary; science was not discussed.

The Sputnik project began in mid-August 1957, and Sergei Korolyov lived at the Tyuratam (Baikonur) missile test range until it was completed. He built a house in a grove of trees that was about halfway between the missile assembly building and the launch pad. He walked to both sites to inspect progress, and held conferences in his small house.[38] He often went to the assembly building and stared at his rocket, brooding over some problem that needed to be solved. Then he briskly rose to his feet, and fired orders to his staff about how to solve the problem.[39]

The satellite itself, the Sputnik, was a simple design. It was a sphere about two feet in diameter, weighing 184 pounds. Its only significant instrumentation was a radio transmitter that announced its presence.[40]

On October 4, 1957, an R-7 rocket stood in the wilderness of Kazakhstan like an elongated pyramid with a primitive, artificial earth satellite as its capstone. Korolyov supervised its launch from a concrete bunker only three hundred feet from the launch pad. At 1930 hours Greenwich mean time (10:30 P.M. Moscow time), the twenty main engines and twelve steering engines of the R-7 booster fired in unison, and the rocket rose smoothly into the night sky on a column of fire several times its own length. Four minutes later it was only a bright star in the northeastern sky. The four tapered boosters burned the last of their fuel, then dropped off. The central rocket, with the Sputnik at its nose, burned for another five minutes until its speed was over 18,000 miles per hour: orbital velocity. An hour and a half after liftoff, *Sputnik I* completed its first elliptical orbit around the earth. Its radio signal repeated a monotonous "beep beep beep" confirming to Sergei Korolyov, his team, and the world that the Soviet Union had put the first artificial satellite into orbit around the earth.[41]

10

Explorer I

We can fire a satellite into orbit 60 days from the moment you give us the green light.

Wernher von Braun[1]

In the evening of October 4, 1957, Wernher von Braun was at the Redstone Arsenal's officer's mess in Huntsville, attending a cocktail party in honor of Defense Secretary-designate Neil McElroy. Other visitors at the party were Secretary of the Army Wilbur Brucker; Chief of Staff of the Army General Lyman L. Lemnitzer; and Chief of Research and Development of the Army Lieutenant General James Gavin. Von Braun was called from the party to take a phone call.

Caller: "*New York Times* calling, Doctor."

von Braun: "Yes?"

N.Y. Times: "Well, what do you think of it?"

von Braun: "Think of what?"

N.Y. Times: "The Russian satellite, the one they just orbited."[1]

Von Braun returned to the gathering to break the news. According to his boss, General Medaris, he began to speak "as if he had been vaccinated by a victrola needle."

"We knew they were going to do it!" von Braun said. "Vanguard will never make it. We have the hardware on the shelf. For God's sake turn us loose and let us do something."[2]

"Sir," von Braun said to Neil McElroy, who was about to become secretary of defense, "when you get back to Washington you'll find that all hell has broken loose. I wish you would keep one thought in mind through all the noise and confusion: We can fire a satellite into orbit 60 days from the moment you give us the green light."

Army Secretary Brucker objected: "Not 60 days."

Von Braun insisted: "Sixty days."

General Medaris, von Braun's boss, settled it: "No, Wernher, ninety days."[1,2]

The following day the ABMA staff at Huntsville briefed Secretary-designate McElroy on its activities with an emphasis on the Jupiter C. Two Jupiter C rockets remained from the Jupiter nose cone reentry studies; and for the past two years had been in storage at Cape Canaveral.[3]

When Secretary of the Army Brucker got back to Washington, he made an even more cautious offer for the Army stating that the ABMA required, "four months from the decision date to place a satellite into orbit," and, he added, "we would require a total of $12,752,000 of non-Army funds for the purpose [a six vehicle program]."[4]

While McElroy and the Washington bureaucracy put off a decision on von Braun's and Medaris's offer, the Soviet Union launched its second artificial satellite, *Sputnik II*, with the dog Laika as passenger on November 3, 1957. The satellite weighed an awesome 1,120 pounds and reached a height of 1,031 miles above the earth. While the first Sputnik sent back only an unearthly beeping radio signal, the second broadcast the heartbeat of its passenger. Anyone who gave it thought quickly realized that the Soviet's next satellite might carry a human passenger.[5]

Five days later on November 8, the Department of Defense issued a press release that read, "The Secretary of Defense today directed the Department of the Army to proceed with launching an Earth satellite using a modified Jupiter-C."[6]

The directive that Medaris and von Braun received from Washington said, however, that they were to *prepare* to launch a satellite. It did not authorize them to light the fuse. Brigadier General Medaris asked his superiors in the Pentagon if their directive meant that the Navy's Vanguard rocket was still first in line on the launch pad, and if the Huntsville team must wait for the final go-ahead. He was told his interpretation of his orders was correct.

After conferring with von Braun, Medaris dictated a telegram to the

Army's chief of Research and Development, Lieutenant General Gavin, stating that if the directive was not changed to give his ABMA group authorization to launch their Jupiter C, he was prepared to resign. Wernher von Braun and Dr. William H. Pickering of the California Institute of Technology's Jet Propulsion Laboratory, which was responsible for building the satellite, were with Medaris at the time, and both insisted that Medaris add their similar intentions to resign to his wire. McElroy and the Pentagon gave in to the threats, and the Jupiter C with the Explorer satellite was on its way to the launch pad.[6,7]

The United States was finally ready for its first shot at space on December 6, 1957. While the Soviet Union had launched its satellites in secret, revealing them only after success, the United States not only allowed press coverage from the start, but made its first attempt to launch a satellite a media event. The media covered it like the World Series or a heavyweight championship fight, which, in a symbolic sense, it was.

The *Vanguard I* stood on the launch pad at Cape Canaveral, fueled and ready to fire. The cameras rolled and news broadcasters commentated as the countdown approached zero. What followed was stamped in the memories of witnesses as a flaming moment of embarrassment and national shame. The *Vanguard*'s rocket-engine fired, the rocket lifted slightly, then settled back to the pad, toppled over, and exploded. The three and one-quarter pound satellite was blasted into the scrub brush and palmetto where it sent back its mournful radio signals.[5,8]

One of the reporters on the scene, Dorothy Kilgallen, remarked, "Why doesn't somebody go out there, find it, and kill it?"[5]

The finger-pointing began before the ashes of the Vanguard were cold.

Democratic National Chairman Paul M. Butler, appearing on the ABC television program "College News Conference," laid the responsibility on the shoulders of Wernher von Braun. He said, playing freely with the facts, that von Braun was "in charge of the program so if there is any responsibility it rests with Dr. von Braun." When asked about a recent comment by von Braun that the United States had done little to develop ballistic missiles in the late 1940s when the Democrats were last in power, Butler said von Braun, who "apparently doesn't remember his own record" never would have come to the United States "if it hadn't been for the vision of President Truman and other members of his administration, who brought him here."[9]

Former President Harry S. Truman, when interviewed at his Independence, Missouri, home said, "He (von Braun) may have been brought over by the Army, but I had nothing to do with it and I never knew him."[9]

Senator Henry M. Jackson (Democrat, Washington) was an inveterate critic of the Eisenhower administration's defense policies. Appearing on the NBC television program *Meet The Press*, Jackson knew exactly where to place the responsibility for the failure of America's rockets and missiles: "It lies with the President of the United States."[9]

Even before attempting to launch their first satellite, America's first satellite, von Braun and his group began to plan subsequent projects. With the blessing of his boss at the ABMA at Redstone, Maj. Gen. John B. Medaris, von Braun and a few of his best people set to work developing a proposal for a "national space program" that would reach at least a dozen years into the future. They ignored interservice rivalries and exploited hardware, boosters, and ICBMs planned and built by all the services. Since apparently only von Braun and his team had been thinking about a space program for over a quarter century, they quickly completed the job and had a written proposal in hand by mid-December 1957. The essential elements of "the von Braun program," as General Medaris called it were as follows:

Spring 1960	2,000 pounds in orbit
Fall 1960	soft lunar landing
Spring 1961	5,000 pound satellite
Spring 1962	circumnavigation of the moon with adequate photographic coverage
Fall 1962	two-man satellite
Spring 1963	20,000 pounds in orbit
Fall 1963	manned expedition to circumnavigate the moon and return to earth
Fall 1965	20-man permanent space station
Spring 1967	3-man lunar expedition
Spring 1971	50-man lunar expedition and permanent outpost

Von Braun and his team estimated that the total cost of the program, which would take fourteen years, would be approximately $21 billion.

The ABMA distributed von Braun's and his team's proposal for a national space program to a select group of Army staff, to their collaborators at the Jet Propulsion Laboratory, and to the Army's chief of ordnance, General Cummings. The Army now took a decisive lead over the other services and the Soviet Union in the planning, if not in the actual implementation, of a space program.[10]

On January 29, 1958, the Jupiter C/*Explorer I* was in place on the launch pad at Cape Canaveral. It was both a complex and an elegantly simple vehicle. It consisted of four stages: The first-stage booster, while designated a Jupiter C rocket, was actually an elongated Redstone; the second stage was a bundle of eleven small solid-fueled rocket motors; the third stage was a cluster of three small rocket motors; and the fourth stage was a single motor topped by the eighteen-pound *Explorer I* satellite. While the Jupiter C booster was controlled by a gyroscopic guidance system, the upper stages were rotated along their axes: They became a gyroscope. The Jupiter C was the product of the ABMA; the upper stages were designed by the California Institute of Technology's Jet Propulsion Laboratory. *Explorer I* carried an instrument for detecting cosmic rays designed by Professor James Van Allen of the State University of Iowa; the instrument would map the distribution of radiation around the earth. Responsibility for launching the rocket—and the *Explorer I* satellite—was in the hands of Dr. Kurt Debus, von Braun's old friend and colleague who had been in charge of test firing the A-4 (V-2) from old Test Stand VII at Peenemuende. While the rocket and the ground support were ready that day, the weather was not cooperative. The jet stream, 36,000 to 40,000 feet above the launch pad, was surging at 170 miles per hour. The launch was postponed. On January 30, the jet stream was blowing at 205 miles per hour. The launch was postponed again.

On January 31, the jet stream was moving at 157 miles per hour, far from ideal, but tolerable. Kurt Debus and his crew began the countdown.

At thirteen minutes before liftoff, the upper stages began to spin at 550 rpm. The engine of the Jupiter C fired at 10:55 P.M. Eastern time, and it rose into the dark sky above Florida. After 157 seconds, when the rocket reached an altitude of 60 miles, the booster engine shut down. Five seconds later, the explosive bolts holding the upper stages fired, and the upper stages coasted upward for another 247 seconds to an altitude of 225 miles. A controller at Cape Canaveral signaled the firing of the second stage, which burned for 6.5 seconds. The third stage burned for an additional 6.5 seconds. Then the fourth-stage engine fired. If all went according to plan, *Explorer I* was traveling at orbital velocity, 18,000 miles per hour.[11]

When the Jupiter C/*Explorer I* vehicle was launched, Wernher von Braun was at the Pentagon in a communications room with Secretary of the Army Wilbur Brucker, Dr. William Pickering of the Jet Propulsion Laboratory, and a handful of generals and top Army scientists. Pickering was on a telephone line to a receiving station in San Diego. According to the flight plan, the *Explorer I* satellite was to cross over San Diego 106 minutes after blast-off.

At that moment, February 1, at 12:41 A.M., eastern standard time, Pickering asked, "Do you hear her?"

"No sir."

Time crept by, seconds, a minute, two minutes.

Pickering on the phone again, "Do you hear her now?"

"No sir."

"Well, why the hell don't you hear anything?"

Secretary of Defense Brucker to von Braun: "Wernher, what happened?"

Soon they were all grumbling and staring at von Braun. "What happened?"

There was no answer, and the generals settled into a silent funk.

Then Pickering, still on the phone let out a whoop. "They hear her Wernher, they hear her!"

Von Braun checked the wall clock and his wristwatch. "She is eight minutes late," he said. "Interesting."[12]

Secretary Brucker shook von Braun's hand, and the generals shook hands all around. Then somebody called President Eisenhower who was in Augusta, Georgia, playing bridge with friends. Eisenhower excused himself from the game and positioned himself at a microphone that was made ready for him to broadcast the news to the world. He said, "The United States has successfully placed a scientific Earth satellite in orbit around the Earth. This is part of our country's participation in the International Geophysical Year."[12] And, although he did not say it, the United States had broken from the starting blocks in the space race.

Besides defending national pride, the *Explorer I* satellite gathered scientific data from space. As it traveled along its eccentric orbit that went as high as 960 miles and as close to earth as 210 miles, it mapped the bands of radiation that encircle the earth and were quickly named after the man who designed the experiment that detected them, the Van Allen radiation belts.[13]

11

Celebrity

The Security of the nation and the free world has been enhanced by his [Wernher von Braun's] great learning and his extraordinary achievements.

President Dwight D. Eisenhower[1]

After the successful orbiting of *Explorer I*, Wernher von Braun became a celebrity as he approached his forty-seventh birthday. He was the darling of the press, and was loved in the halls of power in Washington and—it should not be surprising—in Hollywood. He was not as important as Joe DiMaggio or Marilyn Monroe or Elvis Presley, but he had reached a level of acclaim rarely reached by a scientist. He made the study of science and engineering not only respectable, but exciting. High school boys wrote fan mail to him, and he replied with personal letters and autographed photos.

One question regarding Wernher von Braun's career that is not usually asked is: Why did his employer, the United States Army, let him become a public figure and a celebrity? This question actually goes back as early as December 1946 when von Braun first met the press. Why did the Army let this man, who was in the United States without a visa or passport, speak publicly on a topic that could reasonably be viewed as military research and development? After he had legally entered the country but not yet become a citizen, why did the Army let von Braun participate in the *Collier's* symposium, publish numerous books, act as a consultant to Walt Disney, and appear on Disney's popular television program to discuss rockets? Why did the Army let von Braun become the most recognizable government employee involved in the development of missiles and satellites even before the orbiting of *Explorer I*?

The simple and most obvious reason is that von Braun was good publicity for the Army. He had the sense to remember that the Army was his primary employer and not to let his extracurricular activities seriously impede his job with the Army. Also, he restricted his public discussions to the basics of rocketry, which were public information even if not commonly known, and to speculation about the future of space travel.

The less obvious answer to why von Braun was allowed to meet the public less than two years after he arrived in the United States, must take into account the Army's position within the Armed Forces after the end of World War II. The Army saw its position begin to erode in 1947 with the splitting off of the Air Force as a separate service. The Air Force and the Navy began their own rocket development programs, and the Army was forced to compete with them for funding. The Army, rather than the Navy or Air Force, employed von Braun and his team of rocket experts because the Army had been on the ground in Germany and in a position to exploit the captured personnel and equipment. It helped the Army to have the world's foremost authority on missiles—how else could one view von Braun, the designer of the V-2 and a man who spoke with absolute conviction?—as a visible part of its program.

After the orbiting of *Explorer I*, the competition for funds among the services became intense. Within days of the launch of *Explorer I*, President Eisenhower decreed that "All of the outer space work done within the Defense Department will be done under Secretary McElroy himself."[2] McElroy set up the Advanced Research Projects Agency to manage an emerging, military-controlled space program. Eisenhower and McElroy may have thought they had the military services under control, but the Senate and the services thought otherwise. In 1958, Senate majority leader Lyndon B. Johnson presented his resolution for the establishment of a Senate Special Committee on Astronautical and Space Exploration. The resolution passed, 78 to 1; the lone dissenting vote was from Senator Allen Ellender of Louisiana, who opposed all new committees on principle and was not inclined to make an exception for outer space.[3]

Soon after the successful entry into orbit of *Explorer I*, Capitol Hill was invaded by high-ranking brass and Army press officers pitching their program. They wanted funding to develop a rocket motor that could lift into orbit a fifteen-ton vehicle capable of carrying a passenger. The Air Force, fearing that the Army would get sole responsibility for launching satellites, leaked a story that it was prepared to orbit a 1,000-pound satellite by June

1958 using its Thor intermediate-range ballistic missile as a booster. The Army was soon talking about launching a 1,500-pound satellite.[4]

Before long, Wernher von Braun, the Army's star, was testifying before a congressional committee in support of the Army's proposed extremely powerful rocket engines and, implicitly, asking for a commitment by the government to a massive space program.

> In the field of satellite and moon rockets [at this time, nobody in government except von Braun had begun thinking about a moon project, much less planning one], we need a well planned, long-range national program which makes maximum use of rocket hardware . . . emerging from our ballistic missile and supersonic aircraft programs. This program must be backed up by a firm budget which permits its steady execution over a period of several years.
>
> There is a crying need for more money for basic and applied research in these areas and for development of bigger booster engines.[5]

Von Braun nimbly wove his testimony back and forth across the razor-thin line that separated national defense, as exemplified by ballistic missiles, from national pride, as embodied in bigger booster engines for lifting satellites and exploring the far reaches of space. Von Braun exploited their intertwined needs.

> While adequate research funds are available for clearly defined missile weapons systems, there is never enough to "advance the art." A typical example is that we don't have a really powerful rocket engine today simply because none of our present crash missile programs requires it. But in order to beat the Russians in the race for outer space we absolutely need it—and the development and perfection of such an engine requires several years.[5]

Then he asked for the moon.

> Herein lies the real threat to our security, and only the immediate enactment of a well-planned, determined United States space flight program covering all aspects of unmanned and manned flight through outer space can neutralize it.[5]

The congressmen did not have to be rocket scientists to grasp the issue. The Soviet Union, the Communist juggernaut, was conquering the continents of the earth. Now it was bent on dominating the heavens, too. The United States faced humiliation and the political and physical threat of living under a Soviet sky. The Free World needed a savior, and here before them was the best approximation it had. Von Braun was a prophet of the future, a man who told them that national security and the dream of exploring the heavens were within their grasp, if only they had the courage to reach for it.

The leaders of the nation believed von Braun's message. They would dig deeply into the public's pocket and buy the dream of national security and national pride.

As *Sputnik I* and *II* orbited the earth, crossing the night sky as points of reflected light that everyone on earth could see, the American people began to wonder: How could this have happened? How could the Soviet Union have accomplished something that the people of America knew was critically important to its national survival? Clearly, the Soviets did not have better physical resources and factories; they had been blown into rubble by the Germans during World War II, and were still recovering. They had not done it because their political system was better than ours, because it was an axiom of American political thought that democracy was better than Communism in all ways. Their German rocket scientists were no better than our German rocket experts; and, in fact, there was no reason to believe that any Germans were involved in the launching of the Sputniks. Gradually, the truth sunk in to the American consciousness; their engineers and rocket scientists were better than ours. The educational system of the Soviet Union—at least in the areas of science and technology—was better than that of the United States. Something had to be done about it.

Congress stepped up to the plate and invited Wernher von Braun to do the pitching. In 1958 at the invitation of a congressional committee on education and labor, the Elliott Committee, Von Braun explained how to fix the country's educational system. He was a logical person to ask since he led the team that salvaged the nation's honor by orbiting *Explorer I*.

Von Braun began with a reliable Cold War hardball. He told the Elliott Committee:

> The Communists' threat to the free world extends far beyond armies and politics. It involves every aspect of our way of life: religion, economics, industry, science, technology and education.[6]

The cause of the problem, as exemplified by Sputnik, was that

> the state-controlled educational system is turning out competent engineers and scientists in greater numbers than ours.[6]

Von Braun's general solution was:

> First—we must recruit more young people into scientific and technical careers. Second—we must make these careers more attractive and in-

duce more young people to select them. Third—this involves inspiration, at home as well as in school; competent teachers, adequate laboratories and libraries, assistance to those who require it to finance undergraduate study at least, provision of fellowships or other stipends to encourage graduate study.[6]

Von Braun was savvy enough to advise that any legislation to bolster the educational system would have "very little effect on the guided-missile and satellite programs for the next five years."[7]

Ironically, the big story of von Braun's appearance before the Elliott Committee was not about education but about how he viewed the government's treatment of one of its better-known scientists, J. Robert Oppenheimer, the man who led the team that developed the atomic bomb. In questions that followed von Braun's statement, Representative Frank Thompson Jr., of New Jersey asked about Oppenheimer, who was then director of the Institute for Advanced Study at Princeton, and a resident of Thompson's state. In 1954, after years of service to his country as director of the Los Alamos Laboratory and chair of the General Advisory Committee of the Atomic Energy Commission, Oppenheimer was judged by an AEC panel as loyal to his country, but a security risk. He was then excluded from the government circle he had helped create.

Thompson asked von Braun, "Don't you think it is somewhat tragic that because of political considerations, the country has been deprived of the services of Dr. Oppenheimer?"

"Very much so," von Braun answered. "Particularly the circumstances under which he was dismissed hurt the whole scientific community very badly."

"I agree," Thompson said. "I think it has been an outrage."

"Yes, the way the whole thing was handled [was an outrage]."

While von Braun declined to comment on the security aspects of Oppenheimer's case, he pointed out its absurdity when he said, "I think the British would have knighted him."[7]

Whether or not von Braun had anything to do with it, the federal government poured money into technical and scientific education during the decade that followed, until the Soviet Union was clearly beaten in the space race and other nations became the United States' intellectual and technical rivals.[8]

On May 15, 1958, Sergei Korolyov choreographed the launching of *Sputnik III*, which carried instrumentation to measure gravity and radiation.[9] According to the Soviets, the satellite weighed 2,965 pounds: one and one-half tons.[10] Since *Sputnik III* was sixteen times heavier than *Sputnik I*, it was only logical to conclude—erroneously as it turned out—that the Soviet space program had quickly graduated to a much larger, yet reliable, booster rocket. Korolyov was arguably the Soviet Union's top technical wizard, its equivalent of J. Robert Oppenheimer and Wernher von Braun combined.

In the Soviet Union, it was standard practice to treat engineers and scientists as badly as Oppenheimer had been treated by his country. Under Joseph Stalin, Sergei Korolyov lived and worked in the gulag. Under Nikita Khruschev, Korolyov was condemned to anonymity. The Communist ideology viewed Sputnik as the achievement of the Soviet people and the Communist system.[11] If anyone spoke of Korolyov at all, it was as the chief designer. When he was allowed to speak publicly or write about the space program, it was under the pseudonym Professor K. Sergeyev, who was not identified as the chief designer. He was not allowed to travel outside the Soviet Union where he might talk with others about space exploration. He made his only known trip abroad in mid-1964 to a Czechoslovakian health spa when his health was failing.[12] Inevitably, to the West, Korolyov became a man of mystery, genius, and dread.

Whether he liked it or not—and there is every reason to believe that he liked it—Wernher von Braun was a celebrity. Before the successful launching of *Explorer I*, satellites and the exploration of space were the story. After *Explorer I*, the man behind the success—or at least the man generally perceived to be the cause of it—became the story. *Time* magazine put von Braun on the cover of its February 17, 1958, issue. Other magazines followed.

Time had printed some brief biographical information about von Braun, and would-be biographers scrambled to collect additional facts and anecdotes. To preempt the field and set the tone for all that followed, Wernher von Braun accepted the offer of *The American Weekly*, a magazine inserted into the folds of Sunday papers across the country, to tell his own story. (Before the triumph of *Explorer I*, von Braun had written a piece, "Reminiscences of German Rocketry" for the May 1956, issue of the *Journal of the British Interplanetary Society*, which had limited circulation.)[13]

"Space Man—the Story of My Life," by Wernher von Braun as told to Curtis Mitchell, appeared in *The American Weekly* in three installments on July 20, July 27, and August 3, 1958. Taken at face value, von Braun's autobiography was a simple story of an uncomplicated man with a dream, that of exploring space. He had had the misfortune of being drawn into the sphere of the Nazis and been used by them to subvert his dream to an evil cause. He was, like many Germans, an unwilling passenger on the runaway train set into motion by Hitler. After the defeat of the Nazis, he had come to America to help the Free World and to see his dream live on. Everybody who writes an autobiography naturally presents himself in a favorable light. Many are not beyond including some fictional elements.

Von Braun presented a piece of revisionist history that his biographers uncritically parroted, and became myth.

Many of the details of von Braun's life had already been told; other facts and anecdotes were verified from secondary sources. What is striking in reading his autobiographical article forty years after its original publication, is the twist he gave to anecdotes and the facts he left out. And, of course, there are a few significant misstatements of fact.

Von Braun told the story of how Heinrich Himmler, Reichsfuehrer of the SS and chief of the Gestapo, offered him a position on his staff, how he declined the offer, and was subsequently arrested by the Gestapo, charged with sabotaging the rocket development program by his greater interest in building space rockets. As von Braun told the story, he built credibility for himself as a man of science and as an enemy of the SS and the Nazis. Von Braun also left out the key fact that he held the rank of major in the SS, and that as such Himmler might have had reason to expect some show of loyalty.[13]

Von Braun also wrote that "We heard rumors of concentration camp atrocities. They were unbelievable at first." He made no mention of his first-hand knowledge of the Dora concentration camp, which supplied labor for building the V-2s.

Von Braun reported being congratulated when the V-2 became an operational weapon: "Be happy with your V-2. It's our only weapon that the Allies can't stop. It's hitting London every day."

"It's a success," he claimed to have said, "but we're hitting the wrong planet."

Von Braun wrote in detail about how the Soviet Union took control of the Mittelwerk factory and hired German technicians to run it after the war.[14] He did not mention that members of his rocket team ran it during the war in cooperation with the SS and that they exploited slave labor from the Dora concentration camp.

As von Braun traveled the world, absorbing the glory of the *Explorer I* triumph, he found himself in London in early September 1959. An English reporter asked him, "How do you feel now about your work during the war and its effects on my country?" Von Braun answered, "I greatly regret the abuse of science, but there is an old English saying, 'My country, right or wrong,' and that goes for Germany too." Later in his tour, he was more contrite about the effects of the V-2 on England. "London has always been my favorite city," he said. "I want to say how sorry I am."[15]

The cast of characters in von Braun's life changed as he held center stage. In 1959, Hermann Oberth, his teacher and one of the creators of modern rocketry, returned to Germany, where he and his wife lived on the modest pension he earned as a teacher. He had worked for the Army for only four years, and his advocates in the United States were unable to get him an adequate pension or fellowship that would allow him to stay.[16] On June 2, 1960, Wernher's wife Maria delivered the couple's third child, a son they named Peter Constantine.[17] Those for whom space travel was a fantasy, like Oberth, would soon pass from the stage, while those for whom it would be commonplace were being born.

Celebrity is seductive. If well managed, it can convert transient fame into legend, then into immortality. Nowhere, Wernher von Braun knew, is this transition done better than in the dream factory of Hollywood. Of course, there is the trade-off of strict historical accuracy for dramatic content.

By early 1959, Columbia Pictures was well along in developing a movie based on von Braun's life, which was eventually titled, *I Aim at the Stars*. The movie would be made with von Braun's cooperation and that of his employer, the United States Army. Production of the movie quickly degenerated into a battle between Hollywood's view of artistic creativity and the Army's concept of historical accuracy and its own political agenda. The film's producer, Charles M. Schneer, presented a first draft of the script to Huntsville's Public Information Officer, Gordon L. Harris, for comments on April 14. The screenplay bounced back and forth between the two for some time afterward as Harris made comments on the script that he thought would improve it both technically and dramatically. Schneer had his own thoughts on how the story should be told that did not always follow the Army's line.[18]

Personality conflicts were inevitable. The film's director, J. Lee Thompson, commented that "Von Braun and I disliked each other on sight. After a time I grew almost to like him—almost, but not quite." Thompson was English, and his reaction to von Braun may have been influenced by bad memories, conscious or otherwise, of V-2s falling on London.[19]

I Aim at the Stars lumbered to completion under the burdens of personality conflicts and the mixed blessing of cooperation by the Army. In June 1960, the chief of Army Information hosted a private showing for top government officials including the secretary of defense, the secretary of the army, the army chief of staff, and representatives of the executive branch of government.[20] Senators Lister Hill and John Sparkman of Alabama, with Columbia Pictures Corporation, hosted a preview of the movie based on the life of Alabama's favorite adopted son for members of the House and Senate. The preview was scheduled for 5:30 P.M., Tuesday, June 21, 1960, in the auditorium of the new Senate Office Building.[21]

Cast in the role of Wernher von Braun, was Curt Jurgens, blond, Germanic, dour, and bland. Jurgens and his supporting cast were all taken from the less well-known segment of the Hollywood labor pool: recognizable character actors and actresses, but not top-of-the-line stars.[22]

The lights dimmed and the opening credits rolled. *I Aim at the Stars* told von Braun's life story—or pretended to—from the day he first became interested in rockets through the launch of the first artificial earth satellite. Those who knew von Braun recognized the film's contents for what they were: fiction for dramatic effect. Particularly creative features with no basis in fact were a ludicrous romantic subplot; von Braun's secretary at Peenemuende moonlighting as a British spy; and an American journalist who made a cause out of being von Braun's critic because the rocket builder had worked for the Nazis. Von Braun's real secretary at Peenemuende, who was never a spy, had married one of the Paperclip scientists, and was then living in Huntsville, was particularly upset.[23]

When the lights came up, the audience politely applauded the film and von Braun. Von Braun mingled with the crowd and accepted congratulations at the reception that followed the movie's preview. As he left the reception with his friend and biographer Eric Bergaust von Braun led him to believe that he disliked the film.[23] It was not the story of his life; it was not the myth he had created about his life; it was Hollywood's fiction.

The official premier of *I Aim at the Stars* was held at Loew's Palace Theater in Washington, D.C., three months later on September 28, 1960. It was a fund-raiser, sponsored by the Army Distaff Foundation to create a residence for widows of Army officers. Mamie Eisenhower, who was married to a former Army officer, was there, as was Army Chief of Staff Gen. George H. Decker. Wernher von Braun was also present, making the best of the situation even though he disliked the film.[24]

The critics did not like it either. *Time* magazine was blunt in its evaluation of the film and its artistic merit:

It explodes in a splatter of platitudes about the moral dereliction of the scientific community—personified in von Braun. The movie makers, nervous perhaps about possible public reaction to von Braun's Nazi record and the responsibility he shares for the V-2 attacks on London, have leaned over backwards to stress his war guilt, with the unhappy result that the hero comes off as a jolly accomplice in mass murder, an affable fanatic who cares everything about rockets and nothing about the people they happen to kill. In the end, his moral indifference is shown to be dubiously justified by his scientific success. Von Braun possibly has grounds for a libel suit, but then he might do better to ignore the picture. So might everybody else.[25]

I Aim at the Stars also turned out to be a lightning rod for political protest. Residents of New York City took to the streets in opposition to the film. In London, people who remember the terror von Braun's V-2s brought to the city in 1944 and 1945 also picketed it. In Antwerp, which had been the target of over 1,200 V-2s—more than what fell on all of England[26]—municipal authorities banned the movie outright.[27]

The Army, which still had a massive presence in Germany, threw its support behind the film. A directive from the Army's staff communications office to staff in (Heidelberg) Germany read, "Production of 'I AIM AT THE STARS' was made with full cooperation of the Dept. of the Army. Support of the film is considered desirable to include theater lobby displays, personal appearances of military and civilian officials, and recruiting advertising and publicity."[18] The German people were not impressed. In Munich, the movie elicited boos and catcalls from leftists who were offended by the display of American rockets and missiles.[19]

For adolescent boys who might want more than the flickering memory of a Saturday afternoon matinee, Columbia struck a deal with Dell comic books to do a version of *I Aim at the Stars* as a Dell Movie Classic. The comic book version of von Braun's life, which repeated the fictions of the movie, was uninspiring, self-important, and self-consciously inoffensive— about what one would expect for the ten-cent cover price.[28]

After a short and not very impressive run in theaters, *I Aim at the Stars* mercifully faded from sight. The myth about von Braun's life, which he had so carefully crafted and controlled, had been hijacked by Hollywood and fictionalized. Worse, it had been trivialized. For Wernher von Braun, his film biography was, in the great American tradition of constitutionally guaranteed free expression, a humbling experience.

12

The Challenge of the Moon

We choose to go to the moon in this decade, and do the other things, not because they are easy but because they are hard; because that goal will serve to organize and measure the best of our energies and skills; because that challenge is one that we are willing to accept, one we are unwilling to postpone and one which we intend to win."

President John F. Kennedy[1]

America always does best when it accepts a challenging mission. We invent well under pressure.

Buzz Aldrin[2]

After the success of *Explorer I*, Wernher von Braun became acquainted with the leaders of the United States just as he came to know the leaders of Nazi Germany after the first successful firing of the A-4. Rockets have always needed power—political power—to get off the ground. In Germany, von Braun and his team harnessed the power of the Army, Heinrich Himmler, the SS, and Hitler himself. When rockets and space exploration became national goals of the United States, von Braun counted on the backing of Presidents Dwight Eisenhower, John Kennedy, and Lyndon Johnson.

Although he never truly became an insider, Wernher von Braun developed personal relationships with the men who held the power.

History has painted President Eisenhower as a man with little interest in or understanding of space exploration. His apparent lack of interest in space was based on his twin desires to limit the power of the military-industrial complex, which he warned about as he left office, and to keep the federal budget balanced—both concepts his country declined to accept. While Eisenhower may have been slow to grasp the political significance of space, he caught on quickly and put his indelible imprint on the United States space program. He supported it with balanced amounts of enthusiasm and fiscal caution, and he recognized the leadership and accomplishments of Wernher von Braun.

Wernher von Braun apparently first met Eisenhower at a formal white-tie dinner at the White House on February 4, 1958, held to celebrate the successful launch of *Explorer I*. Guests included William Pickering, James Van Allen, Gen. Curtis LeMay and their wives. Von Braun arrived in Washington with a rented tux, which, he discovered to his dismay as he dressed at his hotel, came with a black tie. He called Eisenhower's press secretary, James Hagerty for help. Hagerty told him that a white tie would be waiting for him at the White House. On arrival at the White House, von Braun adorned himself with a white tie. When President Eisenhower arrived at the dinner, he apologized for his delay. He had been looking all over for his white tie and could not find it. He was somewhat embarrassed at having to wear a black tie instead.[3,4]

Wernher von Braun met Eisenhower again in Washington on January 17, 1959, when the president presented him with the Distinguished Federal Service Medal, the highest award given to civil servants. Von Braun took the opportunity to tell the press, with Eisenhower listening, that his group at ABMA in Huntsville could use another $50 to 60 million to do their job right. For von Braun and his rocket builders there was never enough money. As usual, von Braun was selling, but Eisenhower was not buying that day.[5]

Eisenhower's and his administration's commitment to further ventures in space began in early 1958 when the Advanced Research Project Agency (ARPA) of the Department of Defense met with Wernher von Braun to discuss a possible moon project. As a first step, von Braun suggested building a large booster rocket by clustering engines that were then in production or would soon be available. By mid-summer, on August 15, 1958, ARPA allocated $10 million for a booster rocket based on a cluster of eight Jupiter engines.[6] Previously, on April 2, 1958, President Eisenhower had sent a message to Congress asking for the formation of a civilian space agency. Four months later, on July 29, the National Aeronautical and Space Agency (NASA) was authorized with the already existing National Advisory Committee for Aeronautics as its core organization. On August 20,

Eisenhower assigned manned space flight to NASA.[7] On October 1, 1958, NASA officially came into existence and began operations. One week later, NASA organized Project Mercury. Its goal was to place a manned space capsule into orbit. On January 8, 1959, NASA asked the Army to provide rocket boosters based on the reliable Redstone to lift the first astronauts into suborbital flights.[8]

Von Braun and his group had been in political limbo as Eisenhower choreographed the creation of NASA. Finally, on October 21, 1959, Eisenhower ordered the transfer of the Development Operations Division of ABMA at Redstone to NASA;[5] and the Marshall Space Flight Center became operational on July 1, 1960.[9] The Army lost and NASA gained 4,670 civil service employees and 1,840 acres of Redstone Arsenal land, all placed under the direction of von Braun.[8] In a little less than fifteen months, with the backing of President Eisenhower, Wernher von Braun and his group went from preparing to launch America's first jury-rigged satellite to being at the center of the United States' manned space-flight program.

As the first director of the Marshall Space Flight Center, Wernher von Braun had under his command all of its civil service employees, its land, and its facilities. He ran the operation with the assistance of Dr. Eberhard Rees, deputy for research and development, who had been with him since Peenemuende, and Delmar M. Morris, deputy for administration, formerly of the Atomic Energy Commission. The directors of the ten technical laboratories were Dr. Ernst D. Geissler, Aeroballistics; Dr. Helmut Hoelzer, Computations; Hans Maus, Fabrication and Assembly Engineering; Dr. Walter Haeussermann, Guidance and Control; Dr. Kurt Debus, Missile Firing; Dr. Ernst Stuhlinger, Research Projects; William A. Mrazek, Structures and Mechanisms; Erich W. Neubert, Systems Analysis and Reliability; Hans Hueter, Systems Support Equipment and Director of Agena and Centaur Projects; and Karl Heimburg, Test.[8] Eberhard Rees and every one of his technical directors had been with him since their days together working for Hitler.[10,11]

With the transfer of the Development Operations Division of ABMA at Redstone to NASA, the Army began to clear its files. United States Army Intelligence transferred its records on Wernher von Braun to the security officer at Marshall Space Flight Center. Along with the files came a directive from the assistant chief of staff for Intelligence that read:

> These documents must retain a SECRET classification and cannot be downgraded in the foreseeable future.
>
> The material contains derogatory information relating to one of our most important German scientists who have immigrated to the United States since World War II. This material requires a high degree of protection on the interest of national defense particularly since a U.S. estimate of the security threat involved in this scientist's immigration

is included. The unauthorized disclosure of such information would be particularly damaging to the current Defense Scientists Immigration Program (DEFSIP).[12]

Wernher von Braun's dirty secrets were safe as long as his adopted country still needed him.

On September 8, 1960, President Eisenhower arrived in Huntsville, Alabama, to dedicate NASA's new installation, the George C. Marshall Space Flight Center, named after Eisenhower's World War II commanding officer. When the speeches were done, he joined Wernher von Braun for a tour of the facility, during which von Braun and his staff highlighted the status of their new Saturn I booster. When they reached the static test stand where a Saturn I was undergoing testing, von Braun invited Eisenhower to take the elevator up to the third level for a closer look at the rocket engines. The Secret Service detail, which had inspected the tower, found it at best utilitarian and at worst too creaky for the president's use. Ike looked to the head of the detail for comment. The man in charge slowly shook his head. The president turned to von Braun and said in answer to his invitation to view the rocket up close, "Yes, I believe I would."

The Secret Service watched helplessly as they rode up, as did NASA's public relations people who saw their carefully calculated agenda thrown hopelessly behind schedule.[13]

Photographs of the usually serious Eisenhower taken that day show him in an attentive and often jovial mood.[14] He clearly enjoyed his visit and counted the civilian NASA Marshall Space Flight Center as a victory for his administration.

Eisenhower was in the final year of his two-term presidency as two ambitious men vied to succeed him. Richard Nixon had the mixed blessing of having been Eisenhower's vice president and inheriting the successes and deficiencies of his administration. John Kennedy, with nothing to defend, found failures in Eisenhower's administration and discovered that the United States lagged far behind the Soviet Union in the development of rockets used as weaponry. Kennedy was helped by Eisenhower's secretary of defense Neil McElroy who, in 1959, predicted that by the early 1960s the Soviet Union would have three times as many intercontinental ballistic missiles as the United States. The Democrats in general and Kennedy in particular turned the "missile gap" into a campaign issue.[15] As the presidential campaign ground on, the term missile gap was broadened to include medium- and short-range missiles and expertise in rocketry in general. There is no way of knowing, of course, how many votes the missile gap issue swayed in the election, or if it had any effect on the outcome. Whether or not a gap truly existed is also elusive since the term had quickly lost any specific meaning.

John Kennedy entered office on January 20, 1961. Information that emerged after he took office indicated that the Soviet Union probably did have an advantage in total numbers of operational ballistic missiles. However, the numbers of missiles controlled by both sides was insignificant in relation to the capability of delivering nuclear bombs by long-range missiles. Nevertheless, by early 1961, it appeared that missiles would confer a strategic advantage in the future. Kennedy recognized this fact and was not about to yield the advantage to the Soviet Union. His administration would deliver more missiles and more new rockets.[16]

As President Kennedy was still finding the buttons of power, his Soviet adversaries, Nikita Khruschev and Sergei Korolyov pushed farther into space. On April 12, 1961, Korolyov's *Vostok I* spacecraft carried Maj. Yuri A. Gagarin into space. After one orbit, he landed successfully.[17] However, before Kennedy could do anything to retrieve his nation's lost pride, an unrelated, ill-conceived venture that was already underway took control of the nation's direction.

John Kennedy quickly followed the strategic victory of the 1960 election with the tactical disaster of his equivocal support for the invasion of Cuba by an expatriate army at the Bay of Pigs on April 17, 1961. The invasion, conceived by the CIA, was a series of blunders that ended with the invaders captured or dead.[18] Kennedy swallowed the humiliation by publicly taking responsibility for the fiasco, and swearing to himself to never again be blind-sided by bureaucrats.[19]

Blood still stained the beach at the Bay of Pigs when President Kennedy returned his attention to matters of space. Yuri Gagarin's flight around the earth won for the Soviet Union the prizes of placing first man in space and the first man in orbit. The United States could only be an also-ran unless it defined a grander goal which it had a fighting chance of winning. On April 20, Kennedy sent a five-page memo to Vice President Lyndon Johnson, who Kennedy had appointed to chair the National Space Council, asking for an evaluation of current status and goals in space. Specifically, Kennedy asked, "Do we have a chance of beating the Soviets by putting a laboratory into space, or by a trip around the moon, or by a rocket to land on the moon, or by a rocket to go to the moon and back with a man? Is there any other space program which promises dramatic results in which we could win?"[20]

Lyndon Johnson, the former Senate majority leader, yearned for a mission in the new administration, and seized the initiative on space. He received a memo from Wernher von Braun, presumably in response to his request for comment, stating the view of most senior NASA officials that the Soviet Union's big booster rockets gave it the edge in manned space flight for the next few years. However, if the country were to commit itself to building big boosters, America would have "an excellent chance of beating the Soviets to the first landing of a crew on the moon." This, von Braun added, could be accomplished by 1967 or 1968.[21]

Four days after he received Kennedy's request for an evaluation of the space program, Johnson convened a meeting to which he invited James Webb, Kennedy's appointee as NASA administrator, Jerome Wiesner, Kennedy's science advisor, and several close personal friends whose advice he valued. Johnson also invited representatives of the Army, the Navy, and NASA to make presentations; Wernher von Braun gave NASA's case. After the presentations, Johnson led a discussion, then asked for action.

According to Jerome Wiesner, Kennedy's science advisor, "Johnson went around the room saying, 'We've got a terribly important decision to make: Shall we put a man on the moon?' And everybody said yes. And he said, 'thank you' and reported to the President that the panel said we should put a man on the moon."[20] Both Wiesner and NASA administrator James Webb were bewildered by Johnson's call for action without an unambiguous concensus of the panel, but they could not easily tell the President he should surrender the moon to the Soviet Union.

While Kennedy was coping with the aftermath of the Bay of Pigs, some good news emerged to balance the disasters of the early days of his administration. The results of a Gallup poll released on May 3, 1961, indicated that eighty two percent of the American people supported Kennedy and his administration despite the Cuban fiasco.[22]

Two days later, on May 5, Kennedy and the American people got more good news when Navy Commander Alan B. Shepard became the first American to take a ride into space. An estimated one hundred million people listened to their radios and watched their televisions as the white Redstone rocket, with Shepard tucked into the tiny Mercury capsule at its nose, slowly lifted off the pad at Cape Canaveral.[23] The rocket hurled the Mercury capsule along an arc that took it 115 miles above the earth and 302 miles downrange from its launch point. Fifteen minutes after blast-off, the capsule, with its astronaut, glided back to earth beneath a parachute and landed safely in the sea.[24] The flight was a triumph for Alan Shepard, NASA, the United States, and, of course, Wernher von Braun and his team at Huntsville, who had originally designed the Redstone as an instrument of war.

President John Kennedy addressed the Congress on May 25, 1961, on "urgent national needs." It was, for the President who had stumbled badly out of the starting blocks, his second State of the Union address after only four months in office. The urgent need still remembered by most people, was for a coherent space program.

> I believe that this nation should commit itself to achieving the goal, before this decade is out, of landing a man on the moon and returning him safely to earth. No single space project in this period will be more impressive to mankind or more important for the long-range exploration of space, and none will be so difficult or expensive to accomplish. . . . In a very real sense, it will not be one man going to the moon—if we make this judgment affirmatively, it will be an entire nation. For all of us must work to put him there.[25]

By defining the need for a moon project, Kennedy became the patron of the United States' space program, but he justified it with a weak premise that would inevitably result in the space program's decline. The United States had to show its superiority over the Soviet Union in space since doing so on the battlefield was unthinkable. The purpose of the program was national prestige, not science or exploration. When the moon was reached and prestige in space was in hand, the purpose of the program would no longer be as compelling, and national support would dwindle. But that was not Kennedy's problem in the spring of 1961.

In his second State of the Union address, though generally forgotten, Kennedy presented as "urgent national needs" an expansion of the military—as if it would alter the blunder of the Bay of Pigs—and a comprehensive civil defense program. In Kennedy's words, "We will deter an enemy from making a nuclear attack only if our retaliatory power is so strong and so invulnerable that he knows he would be destroyed by our response." Although Kennedy did not specify the nature of the "retaliatory power," by 1961 the phrase was synonymous with nuclear-armed missiles. Kennedy continued,

> But this deterrent concept assumes rational calculations by rational men. . . . It is on this basis that civil defense can be readily justifiable—as insurance for the civilian population in case of an enemy miscalculation. . . . There is no point in delaying the initiation of a nation-wide long-range program of identifying present fallout shelter capacity and providing shelter in new and existing structures. Such a program would protect millions of people against the hazards of radioactive fallout in the event of large-scale nuclear attack.

As Kennedy saw it, a nuclear war would not be pleasant, but, if America was prepared, it could survive much as England had survived the V-2 bombardment of World War II.[25]

A similar thought seems to have occurred to the leaders of the Soviet Union. Rumors and reports at the time alleged that the Soviet Union had already embarked on a vast fallout shelter program. The confidence of the Soviet people in their fallout shelters is illustrated by a joke that circulated among Russians at the time:

Q. What should I do if a nuclear bomb falls?
A. Cover yourself with a sheet and crawl slowly to the nearest cemetery.
Q. Why slowly?
A. To avoid panic.[26]

Some Americans took the initiative and built fallout shelters for their families in their basements and backyards. Others echoed the sentiment of Soviet cynics and posted instructions for responding to an air-raid siren that read as follows:

1. Go into a crouching position.

2. Place your head between your knees.

3. Kiss your ass goodbye.[27]

It is perhaps indicative of the optimism of the American people that the Congress they elected was far more inclined to spend billions on the grand adventure of travel to the moon than on burrows in which to hide from a suicidal war. In his address, Kennedy had not told Congress how much his space program would cost the nation, but NASA officials who briefed the House Committee on Science and Astronauts gave a ballpark figure of $20 to 40 billion.[28] Congress bought the proposal, and NASA gave birth to the Apollo program.

As the United States planned the construction of a booster rocket to match the capability of the Soviet Union, the Soviet Union struggled to build an ICBM that matched the capability of the United States' weapons with tragic results and far-reaching implications.

While Korolyov's R-7 was a reliable workhorse for launching earth satellites, it had turned out to be a poor ICBM. Only four R-7s were deployed at the Plesetsk military center north of Moscow, and they presented relatively little deterrence to the enemies of the Soviet Union. To fill the gap, the Soviet Union pushed the development of another ICBM designed by one of Korolyov's rivals, an engineer named Mikhail Yangel. Yangel was born in Siberia of German ancestry, and, by one account, during World War II he had infiltrated von Braun's V-2 team at Peenemuende as a Soviet spy. Yangel designed a new ICBM, designated the R-16, which was to be tested at the Baikonor Cosmodrome at Tyuratam under the direction of Field Marshal Mitrofan Nedelin, commander in chief of the Strategic Rocket Forces, which was on par with the army, navy, air force, and air-defense force.

Exactly what happened on the launch pad on Monday, October 24, 1960, may never be fully known, but the most likely scenario has been pieced together by James Oberg, an American scholar of the Soviet space program.

It appeared that late in the afternoon the countdown was halted due to a malfunction. Field Marshal Nedelin, ignoring the most basic safety standards, ordered the launch crew out to the fueled rocket to make repairs, and joined them on the launch pad. The problem arising so late in the countdown made Yangel so nervous that he took the opportunity to step into a fireproof shelter near the launch pad for a cigarette. At about 6:45 P.M., something went terribly wrong. The second stage of the R-16 missile ignited, spewing flame down over the first stage, the gantry and the launch pad. Technicians on the gantry and on the pad near the rocket were instantly turned into screaming torches. The flame quickly burned through the first-stage tanks, and fuel cascaded down to the pad, not exploding in one cataclysmic detonation, but spreading everywhere like a tidal wave of fire. The rising ocean of flame engulfed more personnel who ran in futile attempts to save their lives.[29,30,31]

When the ashes cooled, about 165 incinerated bodies were found on and around the launch pad. Field Marshal Nedelin was among the casualties; the designer Yangel survived.[32] The victims of the R-16 tragedy were buried in secrecy.[31] Field Marshal Nedelin's death was reported as having been the unfortunate result of a plane crash on October 25.[29] The Soviet Union never revealed its failures or tragedies to its people or the world.

The Soviet Union continued flight testing the R-16 through the following year, 1961, but the tests did not go well. By early 1962, the United States, which believed itself to be on the vulnerable side of the missile gap, began deploying squadrons of Atlas ICBMs, thereby putting the Soviet Union at a strategic disadvantage. To restore the balance of power—or terror— Nikita Khruschev considered using Cuba as a base for the Soviet Union's operational ICBMs.[31]

By late 1961, Wernher von Braun, as director of the Marshall Space Flight Center, had become a very powerful bureaucrat, if not a policy maker. He controlled about 40 percent of NASA's budget for the space program. Hugh L. Dryden, deputy administrator of NASA, tried to keep von Braun humble when he thought about the money he controlled. At a 1961 symposium sponsored by the American Rocket Society, Dryden needled von Braun about the nation's lavish spending on space, by saying, "so far all we have obtained for it are some large structures built at Redstone and Cape Canaveral and some noise in the neighborhood of Huntsville."[33]

Von Braun's reply was an exercise in one-upsmanship:

We read a lot nowadays about the need for a crash program in outer space. I think it should also be clearly understood that a space program is a long-range proposition. As Dr. Dryden said, this year's budget is just the down payment for things to be built in the years to come. What we really need in this country is not a crash program but sustained effort over a great number of years. We would be much happier with less of a crash program, but rather a program based on reliable public support of our space program for at least the next decade.[33]

Von Braun asked for support just as John Kennedy asked for a commitment to a national goal to send a man to the moon by the end of the decade. They both got what they wanted.

Wernher von Braun became the man in charge of supplying NASA's big booster rockets, which exploited the technological advances of the preceeding decade. Vehicle structures were now stronger and lighter. The thrust chambers were no longer the solid steel of the V-2 design, but were made of tubular metal. Rocket fuels flowed through the tubes before their injection into the rocket engines; this design resulted in the cooling of the engines and a much lighter construction. Lighter engines and rocket structures translated into bigger payloads. Von Braun and his team at Huntsville incorporated these features into the design of the Saturn I vehicle. The Saturn I's first stage was powered by eight modified Jupiter engines that produced a combined thrust of 1,500,000 pounds. The Saturn I weighed 1,165,000 pounds, stood 188 feet tall, and was capable of hurling into orbit a 22,000 pound payload. It was used primarily as a test vehicle for components of the moon program, but it also launched several research satellites into orbit.

The Saturn I was followed by a similar, but improved rocket identified as the Saturn IB. Its first stage was powered by a cluster of eight improved Jupiter engines that burned liquid oxygen and kerosene to generate a combined thrust of 1,644,000 pounds. The second stage used one of the new J-2 engines that burned liquid hydrogen and liquid oxygen to produce 205,000 pounds of thrust. The Saturn IB weighed 1,300,000 pounds and stood 224 feet high. Although it was powerful enough to put a 37,000-pound payload into orbit, it did not have the power to get a space ship of significant size to the moon.

Wernher von Braun's team at the Marshall Space Flight Center conceived, as the successor to the Saturn I and Saturn IB, the Saturn V, a three-stage rocket that dwarfed its already monstrous predecessor, Saturn I. Saturn V's first stage was powered by five newly designed F-1 engines that burned kerosene and liquid oxygen, each generating 1,500,000 pounds of thrust. Thus, the combined first-stage power of the Saturn V, 7,500,000 pounds of thrust, was five times greater than that of the Saturn I. The second and third stages used the J-2 engine to burn liquid hydrogen and liquid oxygen. As originally designed, the Saturn V stood 364 feet tall and weighed more than 6 million pounds. It was able to put a 140-ton payload into earth orbit or send a forty-seven-ton payload to the moon.[34]

Von Braun and his team at Marshall also had on their drawing boards a rocket design known as Nova, which approached in scale the multistage rocket von Braun had proposed in Collier's a decade earlier. Nova's proposed first stage would have eight F-1 engines generating a total of 6,000 tons of thrust, almost half of the 14,000 tons of thrust of the Collier's vehicle.[35] One engineer remarked that Nova would have weighed so much that "It would have damn near sunk Merritt Island" [the site of launch pads at Cape Canaveral]. Added to the problems of noise and vibration when Nova's first stage fired, the rocket carried with it technical problems and costs that made it impracticable. Nevertheless, Nova contributed to the space program by making the otherwise outrageously large Saturn V seem practicable,[36] and NASA approved a program for developing the Saturn V on January 25, 1962.[37]

In the process of selecting the project director for the Saturn V program, one résumé rose to the top of the pile, that of Arthur Rudolph. His qualifications were exceptional. Rudolph had been a rocketeer for over thirty years; he was a friend of Wernher von Braun's, and he was one of the first to join his team in 1933.[38] He had several years experience as the head of the Development and Fabrication Laboratory at Peenemuende,[39] and two years experience as operations director of the A-4/V-2 plant at the Mittelwerk.[40] Records of Rudolph's activities in wartime Germany were classified, and the ugliness of his collaboration with the SS and his use of slave labor in his factory were long forgotten. Rudolph got the job.[41]

With the rocket-test facility at Cape Canaveral becoming a major NASA base, Kurt Debus, who supervised rocket testing for von Braun at Peenemuende, became its first director in 1962.[41]

By 1962, NASA had a good estimate of the size of the booster rocket it needed to carry a man or men—the number of crew was not yet decided—to the moon. The Saturn V appeared to be the best candidate, though there were innumerable design and performance issues to be worked out. The question that now faced NASA was that of the astrophysical strategy, or in the verbal shorthand NASA adopted, the "mission mode."

A great many possible mission modes were proposed, but after eliminating those that were grossly impractical, three basic scenarios emerged.

The simplest to comprehend and, some said, the simplest to execute was known as "direct ascent." In this mode, NASA would build the biggest rocket booster possible, load on top of it all the hardware needed for the round trip, and blast off. The individual boosters that composed the complete vehicle would drop off one by one as their roles in the mission were completed. Wernher von Braun first proposed a direct ascent mission to the moon in a fictionalized work, "First Men to the Moon," which appeared in *This Week* magazine and later in book form.[42]

The second mission mode, proposed implicit in von Braun's 1952 *Collier's* article,[35] became known as "earth orbit rendezvous," or simply EOR. In von Braun's original plan, a ring-shaped space station would be built first as a site for assembling spacecraft. In EOR, prefabrication of components on earth and a small crew made the space station unnecessary. Components of the moonship would be launched on two or more boosters. They would rendezvous in a low earth orbit, and the crew would assemble the moonship. After assembly, it would blast off with the entire crew to the moon. The advantage EOR seemed to offer was that it required smaller rockets that would be easier to build. The disadvantage that accompanied EOR was that nobody knew how easy or hard it would be to actually rendezvous in orbit. And once rendezvous had been accomplished, there was another new set of problems involved in docking the components and assembling them into a moonship.[43]

The third, and by far the most radical of the three mission modes was "lunar orbit rendezvous," or LOR. The basic concept of lunar orbit rendezvous was developed by NASA engineers at the NASA Langley Research Center in Hampton, Virginia. John Houbolt is generally credited with repeatedly pushing it at NASA senior management until they agreed that LOR might have merit.[44] In the LOR mode, a series of boosters lift the moonship into earth orbit, then send it on a trajectory that places it in orbit around the moon. At that point, an excursion vehicle carrying two astronauts detaches and descends to the surface of the moon. With the lunar landing accomplished, the excursion vehicle blasts off from the moon and rendezvouses with the module it left in lunar orbit. The astronauts then use the rocket capabilities of this module to propel themselves back to earth.

LOR's major advantage was an overall savings in weight that, theoretically, permitted the mission to be accomplished with one Saturn V booster and technology that was already in hand. LOR's apparent disadvantage was that its most complex steps were to be done nearly one quarter of a million miles from earth, and if anything went wrong, there were few things that could be done to fix it, and nothing that could be done from earth. Opponents said LOR was "like putting a guy in an airplane without a parachute and having him make a midair transfer." Proponents of LOR likened it to "having a ship moored in the harbor while a little rowboat leaves it, goes ashore, and comes back again."[44]

The advantages and disadvantages of the three mission modes were, of course, completely theoretical since the spaceships were not yet built and none of maneuvers of the three modes—with the exception of achieving earth orbit and returning—had ever been tried. But each mode had its partisans. Initially, Wernher von Braun favored the direct ascent mode because of its conceptual simplicity and incremental development of hardware.[45] By the end of the 1950s, as technical problems became better understood, von Braun and his team at the Marshall Space Flight Center shifted their support to earth orbit rendezvous, the mode proposed by von Braun in his 1952 *Collier's* article.[35] The newly created Manned Spacecraft Center (MSC) in Houston favored lunar orbit rendezvous. Houston was staffed mostly by American-born men in their twenties and thirties who had no allegiance to Wernher von Braun and who wanted the equipment they were responsible for to do most of the work.[46]

By spring 1962, direct ascent had been discounted by most of NASA's technical people, and EOR and LOR had become the two competing modes. To settle the issue, NASA management in Washington, D.C., assigned the two groups most involved to do technical evaluations of the mode advocated by the other: the Manned Spacecraft Center (MSC) in Houston would evaluate earth orbit rendezvous, and Marshall, under von Braun in Huntsville, would examine lunar orbit rendezvous (LOR). Not surprisingly, the two groups did not reach consensus. To break through the impasse, the MSC-Houston group took their arguments in favor of LOR to Marshall-Huntsville in hopes of selling it to von Braun and his team. The two groups met in the conference room adjoining Wernher von Braun's office on the tenth floor of Marshall Space Flight Center's headquarters building, known locally as the von Braun Hilton. When the presentations were over, von Braun thanked his visitors from Houston and complimented them on their technical evaluation, but he did not immediately endorse their analysis. The Houston engineers were back at Marshall in Huntsville again on June 7, 1962, to hear von Braun's staff's analysis. For six hours, members of von Braun's team gave presentations backing their favored EOR mode. When they had finished, von Braun took the floor.

"Gentlemen," von Braun said, "it's been a very interesting day and I think the work we've done has been extremely good, but now I would like to tell you the position of the Center." To everybody's surprise, von Braun concluded that while earth orbit rendezvous (EOR) was feasible, lunar orbit rendezvous (LOR) "offers the highest confidence factor of successful accomplishment within this decade." Von Braun had decided that it was more important to prevail in the war—to reach the moon as quickly as practicable—than to win the battle over mission mode.[47]

Even though NASA then committed itself to LOR, not everyone in the Kennedy administration fell into line behind it. The President's science advisor, Jerome Wiesner, had from the beginning backed the direct ascent mode, and the exhaustive analyses of the teams at Marshall, Houston, and NASA headquarters had not changed his mind. The disagreement came to a head on September 11, 1962, when President Kennedy and his entourage, which included Vice President Lyndon Johnson, Secretary of Defense Robert McNamara, NASA administrator James Webb, and the President's science advisor Jerome Wiesner, visited Marshall Space Flight Center.

The president and his advisors were in the big vehicle assembly building, dwarfed by the Saturn I vehicle under development, listening to von Braun as he explained LOR with the aid of a large chart.[2]

Kennedy, a man who probed and challenged statements fed to him as indisputable fact, said, "I understand Dr. Wiesner doesn't agree with this." Then he looked around the crowd of bureaucrats, reporters, and rocket scientists for his advisor and said for all to hear, "Where is Jerry?"

Wiesner stepped forward and said, "Yes, that's right. I think the direct mode is better." He began to outline his reasons, and the battle was on. Wiesner went point by point through his list of objections, which made LOR sound poorly planned and based on questionable scientific assumptions. Von Braun, a recent convert to LOR mode, just as forcefully argued the superiority of LOR. The reporters present were listening and taking notes intently; they had their story for the day.[48]

After a few minutes of listening to arguments he probably did not understand, Kennedy put an end to the discussion. To NASA administrator James Webb, he said, "Mr. Webb, you're running NASA—you make the decision." Since Webb had pitched in on von Braun's side in the discussion, his decision was predetermined: lunar orbit rendezvous (LOR).[49]

Before the day was out, von Braun and his staff arranged for Kennedy and his entourage to experience the static firing of the Saturn I. They stood in a shelter 2,000 feet from the test stand when the rocket unleashed 1,500,000 pounds of thrust. For thirty seconds, it poured out fire, smoke, and thunder. Kennedy felt the heat of the rocket on his skin and its power pounding in his chest, and he knew that the moon was within his country's reach.[50]

13

The Cuban Missile Crisis

> The technological revolution has brought about an entirely new set of
> unprecedented problems for the organization of human existence on
> this planet. In fact, most of today's political problems are a direct con-
> sequence of the technological revolution. Capitalism, socialism, and
> communism were unknown words before the invention of the steam
> engine.
>
> Wernher von Braun[1]

The Cold War between the democratic West, tacitly led by the United
States, and the Communist East, under the control of the Soviet Union,
blundered along for a decade and a half until it reached its most potentially
lethal confrontation on the Caribbean island of Cuba. Villainy and stupidity on both sides brought the world to the brink of nuclear warfare and
the incineration of civilization. Wernher von Braun was a primary contrib-
utor to the tools of the crisis, though he was virtually invisible as it took
place. He was also a beneficiary of its outcome.

During the summer of 1962, refugees from Cuba, which had fallen to
Fidel Castro's Communist revolution on January 1, 1959, began telling
stories of military changes in their country. Traffic to Cuban harbors by
ships from the USSR had picked up significantly. The vessels brought many

Soviet civilians, presumably technicians of some kind. Some Cuban refugees reported seeing truck convoys carrying long tubular objects covered by tarpaulins. There was even a report that Castro's personal pilot was heard bragging in a Havana bar that Cuba had long-range missiles with nuclear warheads. The agencies of the United States government discounted these reports since the sources were notoriously unreliable. The Soviet Union had never stationed missiles anywhere beyond its borders, not even in its Eastern European satellites. At that point there was no hard evidence such as photographs.

The concept of the harmless nature of Cuba changed on Sunday, October 14, 1962, when a U-2 photo reconnaissance mission returned from a flight over the western part of the island. The photos, which were developed and analyzed the next day, revealed a trapezoidal shape on the ground that resembled sites that had been seen before only on Soviet soil. Each corner was staked by a surface-to-air missile site (SAM), and the fields themselves contained missile transporters, erectors, and launch pads.[2] On Monday, October 15, the photo analysts also identified Soviet built SS-4 medium-range ballistic missiles (code-named Sandal by NATO) on the ground. According to U.S. intelligence information, the rockets could carry 3,000 pound 3 megaton warheads over a range of 1,100 nautical miles. Within their range were Dallas, New Orleans, Atlanta, and Washington, D.C.[3] As the photo analysts refined their interpretations, the news that within days there would be operational offensive missiles in Cuba worked its way up through the administration.

Special Assistant for National Security McGeorge Bundy, gave the bad news to President Kennedy at 8:00 A.M. on Tuesday, October 16. Later that day, the president summoned key members of his administration to the Cabinet Room. This group became the Executive Committee of the National Security Council (Ex Comm or EXCOM, for short) to advise Kennedy on finding a solution to the threat. At their first meeting the majority saw only one option, an air strike against the missile bases. Kennedy wanted international support for any action and more hard intelligence. The U-2s went back into the air over Cuba.[4]

In the days that followed, the photo reconnaissance interpreters followed the progress of construction of the missile sites and the positioning of the missiles. On October 18, the photo interpreters identified sites capable of launching the SS-5 intermediate-range ballistic missiles (code-named Skean by NATO). The SS-5s were believed to carry 5 megaton warheads and had a range of 2,200 miles. The missiles could reach all of the contiguous forty-eight states, with the exception of parts of northern California, Oregon, and Washington.[5]

The Cuban Missile Crisis was the continuation of a joust between Soviet Premier Nikita Khruschev and President John Kennedy, in which both sides used the ultimate weapons of the times, nuclear-armed ballistic missiles.

Kennedy and Khruschev first met in Vienna in June 1961, not long after the Bay of Pigs fiasco. The major issue they addressed was the presence of United States forces in Europe in general and in Berlin in particular. Khruschev wanted them out. Khruschev, the son of a miner, veteran of the Bolshevik Revolution, survivor of Stalin's purges, and quintessential Communist bureaucrat, was demanding, intransigent, and rude. Kennedy, the son of a rich capitalist and a product of a privileged education, stood his ground throughout and displayed good manners. Kennedy left Vienna feeling shaken—but not beaten; Khruschev left believing Kennedy could be bullied and blackmailed.[6]

In the evening of Monday, October 22, 1962, President Kennedy broke the news of the missile crisis in a televised address to the American people and responded to Khruschev's power play in Cuba.

> Good evening, my fellow citizens. This Government, as promised, has maintained the closest surveillance of the Soviet military buildup on the island of Cuba. Within the past week, unmistakable evidence has established the fact that a series of offensive missile sites is now in preparation on that imprisoned island. The purpose of these bases can be none other than to provide a nuclear strike capability against the Western hemisphere.[7]

Kennedy went on to describe the sites for the SS-4 missiles and their ability to reach the Panama Canal, Mexico City, and Central America as well as Washington, D.C. He said that the SS-5 missiles were capable of reaching targets as far north as Hudson Bay, Canada, and as far south as Lima, Peru. This, he stated clearly, was a threat, not just against the United States but against all of the Western Hemisphere.

Kennedy outlined a seven-step action plan. Its key elements were that the United States Navy would quarantine Cuba, a euphemism for blockade, an action commonly held to be an act of war. The United States would increase its military readiness to deal with the situation and would seek the support of the Organization of American States and the United Nations, and demand that the Soviet Union withdraw its missiles from Cuba.[7]

In the eyes of many American political leaders, Kennedy's program was conservative, weak, and inadequate. Two hours before his public address, Kennedy had met with about twenty Congressional leaders to inform them of the Soviet threat and what his response would be. While they gave him public support, several implied that his approach showed weakness and advocated an immediate air strike.[8] They continued to watch Kennedy for further signs of weakness.

The American people grimly prepared for an unknown future. Sirens announced weekly air-raid drills, though what good that would do if a missile were coming in at thousands of miles per hour was not known.

School children went through their duck-and-cover drills as if the act would save them from immolation, and those few who had built nuclear fallout shelters made sure they were stocked with supplies. To NASA's dismay, the *Huntsville Times* reported that Wernher von Braun was one of those few Americans with the foresight to order a fallout shelter for his family's use.[9]

At first, Khruschev reacted to Kennedy's plan of action with rage, declaring the naval blockade as "banditry, the folly of degenerate imperialism."[10] He had clearly underestimated his rival.

The crisis reached its climax on the morning of Wednesday, October 24. With the blockade lines drawn, twenty Soviet cargo ships approached a fleet of United States warships. They stopped dead in the water, then one by one they turned around. Secretary of State Dean Rusk said to a colleague, "We're eyeball to eyeball and I think the other fellow just blinked." In the next few days, Soviet ship captains allowed their vessels to be boarded and searched before being permitted to pass to Cuba.

Khruschev and Kennedy began to exchange messages by private and official channels, and in moments of deepest concern, by public radio broadcasts. At 6:00 P.M. on Friday, October 26 (1:00 A.M. the following day in Moscow), Kennedy received a letter from Khruschev offering a deal: He would order the dismantling and removal of the missiles if the United States agreed not to attack Cuba. As a return message was drafted, a second letter from Kruschev was received. It contained an additional provision to those stated in the first message. Khruschev wanted the Jupiter missiles stationed in Turkey removed. While it may have seemed to Khruschev and to history a fair exchange, Kennedy could not accept it without appearing to have caved in to Soviet blackmail. At the suggestion of Attorney General Robert Kennedy, the president accepted the proposal of the first message and ignored the second. At 8:05 P.M. on Saturday, October 27, Kennedy sent a message to Khruschev stating his acceptance and pressing for a final agreement by the following day. Rejection would mean invasion of Cuba by the United States.

Just before 9:00 A.M. the following day, Radio Moscow broadcast a message from Khruschev to Kennedy that was also sent through conventional channels. The Soviet premier's message included a scolding for Kennedy's part in the crisis, but also gave his final agreement to the solution.[11]

In the days that followed, President Kennedy ordered low-level reconnaissance flights to monitor the withdrawal of Soviet missiles. He ordered CIA Director John McCone to end all sabotage missions in Cuba. As a final, secret part of the agreement that ended the missile crisis, Robert Kennedy agreed with Soviet Ambassador Dobrynin to remove the Jupiter missiles from Turkey. This would take four or five months to accomplish, and, Robert Kennedy told Dobrynin, if the USSR revealed this concession, the deal was off. President Kennedy was equally forceful with his own generals who resisted removal of the Jupiters. "Those missiles are going to be out of there by April 1," he said, "if we have to shoot them out."[12]

While it was commonly held that the days of the Cuban Missile Crisis were the most dangerous of the Cold War, they were far more dangerous than realized at the time. There were, at the time of the crisis, thirty-six nuclear warheads in Cuba, a number in line with the fears of Kennedy and his advisors. What they did not know at the time was that the local Soviet commander had received authority to use nine of them as tactical weapons in the event of a United States invasion. The escalation into oblivion was almost assured if Kennedy ordered military action.[13] A massive invasion would be met on Cuban soil by short-range nuclear-armed missiles. The United States would be obligated to respond in kind, either against sites in Cuba or the Soviet Union. Then both sides would mount massive nuclear attacks against each other's cities.

Ironically, because of its peaceful conclusion, the two major casualties of the Cuban Missile Crisis may have been Kennedy and Khruschev. Kennedy's handling of the crisis did not endear him to the hawks (factions that wanted a military response), and according to a variety of conspiracy theories, there was involvement by Cubans in his assassination a year later. Nikita Khruschev was deposed on the night of October 14–15, 1964. One of the charges leveled against him was that his arrogance and blundering caused the Cuban Missile Crisis. Fidel Castro, who was virtually invisible during the crisis, was the beneficiary of Kennedy's pledge not to invade Cuba. Although the CIA concocted various plots against him, Castro and his regime outlived both leaders and the demise of the Soviet Union.

The Cold War with its nuclear weapons and ballistic missiles was too dangerous to continue along its natural path without developing some harmless outlet, some benign exercise by which the potential combatants could play out their aggressions without killing everyone. In *The Right Stuff*, Tom Wolfe likened the space race, an appellation acquired soon after the launching of Sputnik, to the ancient practice of single combat. In a tradition that dated to the pre–Christian era, opposing forces picked their mightiest warriors to do battle as an alternative to total war between the two sides. In some contexts, single warriors did battle; in others, small matched groups—for example, three men per side—did battle. Wolfe likened the lobbing of men into space on converted ballistic missiles as a modern incarnation of single combat. The whole world would see who was mightiest, and only the warriors risked their lives. After the Cuban Missile Crisis, the space race became the surrogate for nuclear war.[14]

The warriors of both the Soviet Union and the United States rode into battle on steeds that traced their lineage to Wernher von Braun's V-2. Two astronauts made America's first brief flights into space as the payloads of the old reliable Redstone missile. While the Mercury orbital flights were powered by the Atlas ICBM and the Gemini flights would be boosted by a Titan 2, von Braun once more oversaw the design and construction of the booster for the Apollo flights to the moon. And as America's leading rocketeer, he would participate in plotting strategy for the coming battle.[15]

Even as the expeditions into space and the race for the moon pressed on, fear of annihilation by nuclear warfare persisted in many minds. In 1964, a little over a year after the Cuban Missile Crisis, the movie *Dr. Strangelove*, directed by Stanley Kubrick, portrayed the apocalypse as the darkest of black comedy. The chilling plot showed strategic stupidity and tactical blunders bringing the great powers to nuclear war. Peter Sellers played the title role of the shadowy, brilliant, but amoral wheelchair-bound scientist who is the president's Director of Weapons Research and Development. Strangelove's crippled body and soul betray him when his right arm uncontrollably snaps into a Nazi "Sieg Heil" salute or grabs his throat in suicidal self-strangulation. With his smooth German accent, he euphorically addresses the president as "Mein Fuehrer."

As the enormous map at the rear of the War Room tracks the paths of both Soviet and American ballistic missiles, Dr. Strangelove proposes an alternative to the annihilation of humanity. "Mr. President?" he says in his high-pitched voice, "I would not rule out a chance to preserve a small nucleus of human specimens. It would be quite easy . . . at the bottom of some of our deeper mine shafts." Strangelove suggests an ironic reproduction of the Mittlewerk, built decades earlier in a German mine where a city of slaves built the V-2 missiles.[16]

Years after the movie's release, many saw in Dr. Strangelove the image of Henry Kissinger, who did not appear on the political scene as President Nixon's National Security Advisor until 1968, four years after the movie debuted. The movie's creator, Stanley Kubrick, found the resemblance to Kissinger unintended and an unfortunate coincidence.[17] In 1964, the one ex-Nazi, German scientist who had the ear of presidents, who was the logical model for Strangelove, was Wernher von Braun.

Wernher von Braun greeted one of his most important supporters, President John F. Kennedy, when he arrived at the Marshall Space Flight Center on September 12, 1962. Eclipsed behind Kennedy's left shoulder is an even more enthusiastic supporter of the space program, Vice President Lyndon Johnson. Standing prominently between Kennedy and von Braun is Alabama Congressman Robert Jones. Photograph courtesy of NASA, Marshall Space Flight Center.

Walt Disney, constantly in search of new ideas for his movie productions and entertainment parks, visited Wernher von Braun at the Marshall Space Flight Center on April 13, 1965. Walt Disney is at the left and von Braun is at the far right. Photograph courtesy of NASA, Marshall Space Flight Center.

Wernher von Braun, photographed in his Marshall Space Flight Center office with rocket models, most of which were designed by his team. Photograph courtesy of NASA, Marshall Space Flight Center.

A Saturn IB booster carries *Apollo 7* with astronauts Walter M. Schirra Jr., Donn F. Eisele, and Walter Cunningham into earth orbit on October 11, 1968. Photograph courtesy of NASA, photo no. 68-H-930.

One month before their scheduled flight to the moon, *Apollo 11* astronauts (from left to right) Michael Collins, Neil Armstrong, and Edwin (Buzz) Aldrin pose before the Lunar Landing Module Simulator. Photograph courtesy of NASA, photo no. 69-H-832.

The Saturn V/*Apollo 11* space vehicle carrying astronauts Neil Armstrong, Michael Collins, and Edwin (Buzz) Aldrin lifts off at 9:32 A.M. EDT, July 16, 1969. Photograph courtesy of NASA, photo no. 69-H-1142.

Apollo 11 officials are in a festive mood after its successful liftoff on July 16, 1969. They are, from left to right: Charles W. Matthews, deputy associate administrator for Manned Space Flight; Wernher von Braun, director of Marshall Space Flight Center; George E. Mueller, associate administrator for Manned Space Flight; and Lt. Gen. Samuel C. Phillips, director of the Apollo Program. Photograph courtesy of NASA, photo no. KSC-69P-641.

The *Apollo 11* astronauts photographed the moon from a distance of 10,000 nautical miles. Photograph courtesy of NASA, photo no. 69-H-1374.

Wernher von Braun, having moved on to become NASA associate administrator for Future Programs, kept track of the Apollo program by attending the launch of *Apollo 14* from Kennedy Space Center on January 31, 1971. Photograph courtesy of NASA, photo no. 71-H-253.

A Saturn V booster rocket was used to launch the United States' first and only space station, Skylab, on May 14, 1973. Photograph courtesy of NASA, photo no. 74-H-98.

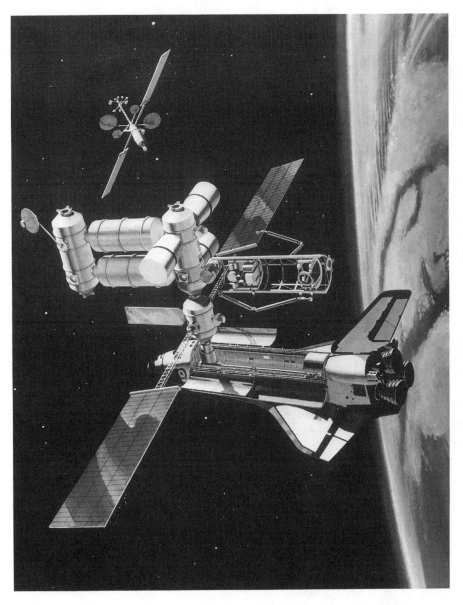

In this artist's concept, the space shuttle docks with a space station. The fundamental features of the space shuttle were defined during Wernher von Braun's tenure as NASA associate administrator for Future Programs. Photograph courtesy of NASA, photo no. 82-H-869.

Having sent men to explore the moon, NASA puzzled over what to do next. This artist's concept from the 1970s illustrated a proposal to build a mining colony that was reminiscent of the lunar exploration base proposed by Wernher von Braun in the October 25, 1952, issue of *Collier's*. Photograph courtesy of NASA, photo no. 77-H-265.

In his final year, though in declining health, Wernher von Braun kept his optimism and enthusiasm for life. Photograph courtesy of NASA, photo no. 77-H-39.

14

The Moonship

There's no more problem to building a "bigger" rocket than there is to building a "bigger" airplane. It just costs more money.

Wernher von Braun[1]

The journey of the United States to the moon began in January 1960, when President Eisenhower directed NASA to accelerate development of a "super booster," when NASA began planning a multiman space vehicle, and when NASA senior staff began using the name Apollo to describe the spacecraft that would go to the moon.[2] President Kennedy's endorsement of a lunar mission as an urgent national goal in 1961 gave the project further momentum. The consensus in NASA of lunar orbit rendezvous (LOR) as the mission mode inevitably defined the major components of the moonship. All that remained was designing and testing the millions of parts of the moonship that had to work without failure if men were to go to the moon and return safely.

By 1962, the basic design of the moonship was worked out. At its base was the Saturn V super booster, which was composed of three stages: The first stage was the Saturn S-1C, which had five F-1 engines that burned kerosene and liquid oxygen and delivered a total of 7,500,000 pounds of thrust. The second stage was the S-II, which burned liquid hydrogen and liquid oxygen in its five J-2 engines. The third stage was eventually designated the S-IVB, and it also burned liquid hydrogen and liquid oxygen in one J-2 engine. Above the Saturn V were those components intended to function in space and on the moon: the Lunar Module, which would descend to the surface of the moon; the Service Module, which contained the fuel and rocket engine needed for the earth-to-moon and return flight; and the Command Module, which would be home for three astronauts.[3]

The sheer complexity of the moonship required that its design and construction be delegated to more than one NASA center. Design and construction of the Saturn V was the responsibility of Wernher von Braun and his group at the Marshall Space Flight Center. Design and construction of the Command and Service Modules and the Lunar Module were largely under the control of the Manned Spacecraft Center in Houston. NASA contracted construction of the first stage (the S-IC) to the Boeing Company, the second stage (the S-II) to North American Aviation, and the third stage (the S-IVB) to the Douglas Aircraft Company.[4]

NASA had become a bureaucracy with several centers and layers of management responsible for the Apollo mission. Into this labyrinth came George Mueller, new head of the Office of Manned Space Flight and a man who, by clarity of vision and personal will, brought the goal of a manned lunar landing by the end of the decade within reach. Mueller had worked his way up to the NASA position from the Air Force's ballistic missile programs, where the final result was more important than the elegance of the approach. When he arrived in September 1963, one of his first actions was to order a brutally honest evaluation of the Apollo program with a realistic evaluation of when the first lunar landing could be made. At the time, the best estimate was that the first attempt at a lunar landing would be made in late 1971. Mueller clearly saw that the current program plan was unacceptable and that a radical change had to be made to accelerate it.[5]

Mueller presented his solution to the problem, not as a proposal, but as an edict. On October 29, less than two months after he took his new job, he presented the management teams from Huntsville and Houston with the concept of "all-up testing" and a test-flight schedule based on the concept that would keep to President Kennedy's end-of-the-decade target.[6] Stated simply, a stage of a rocket vehicle was considered "up" if it was ready for flight testing. "All-up" meant that all stages of a spacecraft, the Saturn V, to be precise, would be readied for flight and tested at the same time. With all-up testing, there would be far fewer test flights, and less time would be needed to get the first manned expedition on the launchpad.

To Wernher von Braun and his crew at the Marshall Space Flight Center, this approach appeared to fall somewhere between heresy and insanity. Ever since their early days in Germany, even before Peenemuende, von

Braun and his team had progressed incrementally, building and testing each component before moving on to the next. There had been about 3,000 test firings of various versions of the A-4/V-2, most of them before the missile went into operation. For the mission to the moon, they had planned a shake-out of each stage of the Apollo/Saturn V vehicle separately before joining them in combined sequential firings, as they were ultimately intended to function. Mueller relentlessly argued (though he had already dictated his conclusion as the plan) that if one were to adequately test the first stage, it must carry a payload that approximated the upper stages. And if von Braun and his team were to go to the trouble of building dummy upper stages, they might just as well build and fire the real thing. Besides, Mueller pointed out, there was no way to realistically test the upper stages with ground launches. The upper stages were intended to be fired when they were already in the upper atmosphere or higher, traveling at speeds of thousands of miles per hour. Mueller's most telling argument for all-up testing, however, had nothing to do with technical matters. He simply made it clear that without taking the apparent shortcut, there was no way to get to the moon by the end of the decade. And if that target date was not in the project plan, Apollo might not fly at all.[6,7] Von Braun and his team gave in just as they had a year earlier when they agreed to the concept of the lunar orbit rendezvous (LOR) mission mode.[6]

Two-and-a-half years had passed since President Kennedy presented a manned flight to the moon as an urgent national priority, yet he had kept up his enthusiasm for the Apollo program. On Saturday, November 16, 1963, he arrived at the Launch Operations Center at Cape Canaveral to see how NASA was progressing on the project. Wernher von Braun was there to show the president the Saturn I and mock-ups of the hardware for the Apollo/Saturn spaceship.[8] When he finished with von Braun, Kennedy joined Kurt Debus, von Braun's friend and colleague and the newly appointed head of the center, for a helicopter tour of the moon base.[9] Significantly, the helicopter also carried two Americans who had already flown into space, Gus Grissom and Gordon Cooper, who, with pride and excitement, told the president about Apollo.[8] The astronauts were already beginning to take the place of von Braun and his German team as the center of attention in the American space program.

Kennedy was apparently impressed by what he saw at the Launch Operations Center at Cape Canaveral. Congress was not. On the following Wednesday, November 20, the Senate cut $612 million from Kennedy's budget request for NASA. A man on the moon by the end of the decade was too remote, too far beyond the next election. Kennedy was actively pushing his moon program the very next day as he began a tour of Texas to reunite the Democratic political party. In a speech in San Antonio, he praised the space program. Later that day in Houston, the new home of the Manned Spacecraft Center, he attended a testimonial dinner for a local man who was instrumental in bringing the center to that city. The following day, November 22, Kennedy rode in an open car in a motorcade in Dallas. At 1:00 P.M., he was pronounced dead, the victim of an assassin's bullet. Six days later, NASA renamed the Launch Operations Center the John F. Kennedy Space Center.[9]

The reaction of some in NASA to the assassination of the president was that the bullet that killed President Kennedy also meant the death of the Apollo manned moon program.[10] Lyndon B. Johnson, Kennedy's successor as president, thought otherwise. Johnson gamely embraced Kennedy's programs and was an even greater supporter of space exploration. It was he, after all, who had sold Kennedy on the idea of sending a man to the moon by the end of the decade.

By the end of its first year of existence, the Marshall Space Flight Center under Wernher von Braun's direction employed 5,500 people, and the number rose to 7,500 at the peak of Apollo activity in 1966. Marshall's budget, which von Braun managed, was even more impressive; it surpassed $1.5 billion per year for the peak years 1964 through 1967.[11] The vast majority of the money was not spent in Huntsville, however. NASA had committed over 90 percent of Apollo funding for the building of facilities and hardware, and for the operation of some NASA facilities.[12] The overall Apollo program eventually required the direct involvement of about 20,000 industrial and university contractors and the participation of 400,000 people.[13]

One great irony of the Apollo program was that it turned Wernher von Braun and the members of his original team from a group of hands-on rocket builders into contract managers. They were the experts who were charged by NASA with directing and supervising the contractors, primarily the giants of the aerospace industry. Von Braun and his team fought the tide that would raise them above the creative work. In 1963, Marshall underwent a major reorganization intended to relieve von Braun and other Marshall top management officials of day-to-day management of the greatly expanded operations and permit them to focus on policy matters and major engineering considerations. While they might have lost the organizational battle, they succeeded brilliantly as project managers.[11]

As expansive as the space program was in the mid-1960s, it competed with other foreign and domestic issues for the attention of the American people. The Johnson administration was leading the country into disastrous involvement in the civil war in Southeast Asia on the side of the corrupt, autocratic South Vietnamese government against the Communist Vietcong, who were supported by North Vietnam. At home, the United States confronted the legacy of its own shameful legacy of slavery: racial segregation.

To his credit, President Lyndon Johnson orchestrated the passage of and signed the Civil Rights Act of 1964.[14] Less patient civil-rights activists pressed the issue of voter registration in the South. The ideological conflict came to its climax on "black Sunday," March 7, 1965, in Selma, Alabama. A group of about 600 activists—mostly black—planned to march from Selma to Montgomery to initiate a voter-registration drive. Alabama Governor George Wallace, who had previously distinguished himself by barring the doorway of the University of Alabama to racial integration, declared the march a menace to commerce and public safety, and he sent 200 state troopers to back up the local sheriff. As the marchers crossed the bridge over the Alabama River, the sheriff's deputies on horseback and the troopers routed the marchers with tear gas, clubs, and bullwhips. Television cameramen and reporters covering the march insured that the country witnessed Selma's shame. Two days later, Dr. Martin Luther King Jr. led a second, symbolic—and peaceful—crossing of the Selma bridge, and under pressure from black activists, he became involved in the planning of a new march to Montgomery.[15]

Under pressure from the federal government and with the participation of federal agents, the march from Selma to Montgomery was made on March 21 to the jeering of unreconstructed rednecks and the intimidation of Ku Klux Klansmen. In the shadows beyond public view, the Selma demonstrations were accompanied—in separate incidents—by the murders of three civil-rights activists. To the nation and the world, the white citizens of Alabama, its public officials, and its governor appeared to be vicious segregationists with at best transient respect for the law and none for the rights of blacks.[16] Alabama and Governor George Wallace took a much deserved battering of their images, and Wallace looked around his state for something to rehabilitate it, something racially neutral, nationally supported, and undeniably progressive. He found what he was looking for 160 miles north of Selma at Huntsville.

By 1965 Wernher von Braun's first spaceship had been on its launching pad, and the Rocket to the Moon ride had taken off with dozens of passengers several times each hour for ten years. The world's first spaceport at Disneyland in Anaheim, California, had gone beyond success to become an institution. By the mid-sixties, however, its creator, Walt Disney was gathering ideas for his next entertainment center. Tomorrowland had been a big hit, but it was being overtaken by the reality that had been created by Disney's partner in its creation, Wernher von Braun.

Disney, with an entourage of six, spent Monday, April 12, 1965, at the Manned Spacecraft Center in Houston and then flew to Huntsville later the same day. They arrived at the Huntsville Municipal Airport in their corporate jet in the early evening. Disney arrived in time to be an honored guest at a dinner meeting of the North Alabama Section of Rotary International, at which his old friend and colleague Wernher von Braun was the featured speaker.

The following morning, Disney and his party met von Braun at his office at the Marshall Space Flight Center. Von Braun gave them a briefing that included an illustrative slideshow about all he had helped create in the past ten years: the Apollo mission profile, the Marshall Space Flight Center, and the Saturn rockets. Von Braun spent most of the day showing them the major laboratories at Marshall, the mock-up of the Saturn booster, and the Saturn static-test facilities. At the end of the day, an educated and impressed Walt Disney departed with his associates on their jet for the Cocoa/Titusville Airport. The following morning they toured the Kennedy Space Center.[17]

At the time, Disney's company was secretly buying up property in central Florida, about sixty miles west of the Kennedy Space Center. By October 1965, the company had acquired 27,443 acres, nearly forty-three square miles, an area roughly twice the size of Manhattan Island. At a press conference on November 15, 1965, Disney, accompanied by the governor of Florida, announced that this land would be the site of his new entertainment center, Walt Disney World.[18]

Quite possibly, von Braun was more inspired by his meeting with his former collaborator than was Disney. Even before von Braun and his team put the *Explorer I* satellite into orbit, Redstone Arsenal had been collecting obsolete rocket hardware, including the V-1 and V-2, in what became an ad hoc museum of rocketry. When the Marshall Space Flight Center was spun-off under von Braun's direction, it, too, started a collection of outdated equipment and prototypes. During the early 1960s, von Braun advocated the creation of an exhibit of the merged collections, which would be open to the public, at a site on neither the Redstone nor Marshall bases. In scale, scope, and ambition, it would be what Walt Disney might create if he had the contacts and money.

In early June 1965, two months after Walt Disney toured the Marshall Space Flight Center, three months after the civil-rights demonstrations, the blundering official repression, and the racially motivated murders at Selma, Governor George Wallace arrived at Marshall for the VIP tour and requisite photo opportunity. Marshall released a publicity photo of Wallace, Wernher von Braun, and NASA Administrator James Webb; all were smiling.[11] Wallace was no doubt pleased that his appearance at the rocket-building capital of the Free World took some of the dents out of his battered public image and gave it a progressive paint job. Webb had reason to smile because his agency operated one of the few, and possibly the only, federal activity in Alabama that was appreciated by its people and governor. Von Braun was pleased because he acquired his most important ally in the creation of his pet project, the rocket and space exhibit at Huntsville. By September 1965, the Alabama state legislature had approved a $1.9 million bond issue to fund the enterprise. Governor George Wallace, warming to the role of advocate of science, technology, and education, appointed von Braun to a seat on the new Alabama Space Science Exhibit Commission.[19] The United States government helped the project along by deeding thirty-five acres to the State of Alabama for the site.[20]

15

Blunders and Disasters

> Early in the planning phase it had been recognized that . . . there were too many things to do, too many things to observe, and too many things that could go wrong from the tiniest oversight.
>
> Wernher von Braun[1]

As the Mercury program ended, Gemini carried men into orbit, and the Apollo juggernaut moved relentlessly toward the moon, the men—and they were all men—who became synonymous with space exploration were mostly astronauts. Those at the top in NASA at its Washington headquarters, Administrator James Webb; Deputy Administrator Hugh Dryden; head of the Office of Manned Space Flight George Mueller; and Gen. Sam Phillips who headed the Apollo Program Office, were all faceless bureaucrats.

The name of only one earthbound player was generally recognized, that of Wernher von Braun. He accepted his new role, far from obscurity but no longer center-stage, with good humor. In fact, von Braun got along very well with the astronauts and had their respect and admiration. This may have been because they all had the same dream and the astronauts were mostly, like von Braun, engineers. They spoke the same language.

As the astronauts were preparing to take over the making of history, von

Braun was consolidating his position in history as a scholar and a man of public affairs. Since the successful launch of the *Explorer I* satellite in 1958, he had been collecting awards from civic and professional groups the way another man might collect bowling trophies. By the end of 1966, he had accepted nineteen honorary doctorate degrees, and his résumé in 1966 listed memberships, active and honorary, in eighteen professional organizations.[2,3] Noticeably absent from the list was membership in the American Association for the Advancement of Science (AAAS), which had invited von Braun to join in 1958. The AAAS was the largest and most diverse scientific organization in America. Von Braun declined the invitation on the advice of Army security, which was still guided by McCarthy era paranoia and perceived Communist influences in the AAAS's advocacy of civil liberties and other liberal causes.[4]

Fame begets fame, and von Braun took advantage of every opportunity to keep his name and ideas in public view. After the success of *Explorer I*, he was besieged with invitations to write magazine articles on rockets and space travel. A bibliography of his writings through 1969 lists several hundred pieces. Even allowing for multiple publications of the same item in various languages, his output was prodigious. Surprisingly, von Braun's bibliography listed very few technical and scholarly papers, and those he did write were technical reviews or speculative pieces.[5] Von Braun was clearly a popularizer who had little interest in the discipline of peer review of his scientific and technological work.

In 1963, he signed on for a regular writing assignment with *Popular Science* magazine. He produced monthly articles for the next ten years, and bimonthly pieces thereafter.[3] The level of technical sophistication of his writing is indicated by the first two articles: "Wernher von Braun Answers Your Questions"[6] and "More Answers to Your Questions About Space."[7] Two articles from 1965 reveal a not-so-latent apprehension about the long-term health of the space program: "When Will We Land on Mars?"[8] and "Whatever Happened to the Manned Space Station?"[9]

In 1966, von Braun coauthored the exhaustive and masterfully illustrated *History of Rocketry and Space Travel*.[10] His coauthor and friend, Frederick I. Ordway III, would write other pieces with him and join the choir of his uncritical and reverential biographers.

Wernher von Braun's work for NASA and his numerous extracurricular activities dictated a need for an efficient, loyal support staff. He relied on a second team, distinct from his German-born rocket development team, to keep his professional and personal publicity operation going at Marshall Space Flight Center.

Eberhard Rees was deputy director of Marshall, and for thirty years he had been von Braun's deputy and confidant. He lacked the personal flash of his boss, but he was an expert at getting the day-to-day job done. He would succeed von Braun as director of Marshall in 1970.[11]

Bonnie Holmes was von Braun's secretary. By all accounts, she was a superb office manager with tremendous loyalty to von Braun.[12]

Bart J. Slattery Jr. was chief of information, Public Information Office of the Marshall Space Flight Center. Slattery was an ex-navy captain, who ran interference between von Braun and anyone who might be inclined to write critically about the space program or the director. Like most political figures of his day—and those who followed—von Braun also had two ghost writers who put words to his ideas for many of his speeches. However, the characteristically simple language, dearth of jargon, and reliance on American colloquialisms revealed von Braun's controlling hand in the writing of his speeches.[13] It is not known to what extent von Braun's ghost writers also contributed to his prodigious output of magazine articles and books.

Since Wernher von Braun became a celebrity, part of his job was to remain one. He gave interviews, the contents of which were preordained. The press was, as Tom Wolfe described it in *The Right Stuff*, "the consummate hypocritical Victorian gent."[14] They resisted asking embarrassing questions and held back inconvenient information. The astronauts were all stable family men; there were no reports of their wild escapades. Likewise, they held back on Wernher von Braun. The press treated him as the prophet of the space age and as a minor philosopher and theologian of the mid-twentieth century. They pitched him soft-ball questions that were sure hits. No one asked about what he had done in Germany before 1945. Anyone who dared would be cut off from future access. Not surprisingly, printed interviews with von Braun were almost uniformly marginally informative and almost always boring. Reporters almost always got the official public image.

One reporter who interviewed von Braun and failed to fit the profile of the Victorian gent was Oriana Fallaci. She wrote about herself and birthplace, "I grew up in an anti-Fascist family, and during the Nazi occupation of Florence, I fought in the Underground Movement: Corps of Volunteers for Freedom. At fourteen I was honorably discharged by the Italian Army as a simple soldier."[15] In the mid-1960s, she was living in the United States while researching a book on the scientists and astronauts of the American space program. Her interview with von Braun was de rigueur.

On her arrival at Marshall Space Flight Center, Fallaci was taken to a conference room with a long table surrounded by chairs. She waited alone for her subject.

The interview got off to a fast start.

"I cannot apologize enough. I am eleven minutes late."

"It doesn't matter."

"It does matter. And I'm sorry, because I cannot spare you more than half an hour. I am never late."

"I know."

"You know?"

Von Braun took off his raincoat, tossed it into a chair.

"My name is Wernher von Braun."

"I know."

"What is yours?" There was little doubt who was in charge of the interview.

Von Braun engaged her in small talk—as what was left of her half hour ticked away—and sat at the conference table taking the chairman's place. As he did, Bart Slattery, Marshall's chief of information, hurried into the room, late and breathless. Fallaci described him as "a little man with a red obsequious face and stooping obsequious shoulders." Slattery tried to take charge of the interview. Von Braun brushed him off. He was in charge.

Time trickled away. Fallaci began. "Mr. von Braun, I'll omit the preliminaries and put a question to you at once. The question is this—"

Bart Slattery slid a slip of paper across the table to Fallaci. On it was the message, "*Doctor* von Braun. *Not* Mr. von Braun."

Fallaci, a bit confused, blundered on. "The question is this, Mr. von Braun. Here people talk of the journey to the Moon as they do of the trip from Huntsville to New York and repeat that, for America at least, it will take place by 1970—"

Slattery slid another note across the table to the reporter. "DOCTOR von Braun!!!"

"Will it really take place by 1970, *Doctor* von Braun?"

Slattery nodded happily that she finally got it right, and von Braun shifted his attention from his fingernails to the reporter.

"Provided the people of America are willing to pay—yes, there's no doubt of it. . . . Obviously there are certain difficulties, but all perfectly surmountable. It's only a short journey; eight days there and back. Going to the moon is a picnic."

"A picnic?"

"A picnic, a trifle, a party trick."[16]

True, by the time of the interview, von Braun and his team at Marshall Space Flight Center had built the Saturn I, and it had been successfully flown from the Kennedy Space Center. But the much larger Saturn V, the Apollo spacecraft, and the lunar lander had yet to get off the ground.

The journey to the moon was von Braun's party trick. Von Braun guided the interview along a more ambitious course: Mars, eighteen months round trip; Venus; the Russians; and God.

Fallaci, like those before her, forgot the V-2 and the Nazis. She was sucked into von Braun's whirlwind of enthusiasm.

"I see ethical principles there," he said. "Two factors are necessary to make man accept ethics: one is the belief in the Last Judgment, when every one of us must account to God for his use on Earth of the precious gift of life, the other is the belief in immortality, which is to say the continuation after death of our spiritual existence. Because we have a soul." Fallaci's mind drifted back to the war and the enemies of the Nazis being shipped off to Germany, never to return.

"I hope I've made myself clear."

"Perfectly clear, Dr. von Braun. Perfectly clear."

"That's thirty eight minutes," Bart Slattery interrupted, "eight more than the allotted time."

Von Braun said, "I have to go now."

He was a busy man, an important man. He left the conference room as abruptly as he had arrived. Von Braun's curt style and his authoritative air brought back to Fallaci unpleasant memories of the days when jackboots sounded on the streets of Florence.[16]

While von Braun's carefully constructed image attracted an occasional challenge such as Fallaci's in the United States, it took a pounding in Europe. In 1963, an East German named Julius Mader published a book, *The Secret of Huntsville: The True Career of the Rocket Baron Wernher von Braun*. The tone of the book was clearly indicated by its cover, which was dominated by a drawing of von Braun wearing an SS uniform complete with death's head emblem on its peaked hat. Madder's book correctly reported that von Braun held the rank of major in the SS and that the V-2 was produced by slave labor at the Dora-Mittelwerk facility. Madder also reported that Dora prisoners had seen von Braun there in SS uniform on several occasions. Regrettably, Madder's biography about von Braun was a product of the Cold War. It was strident in its criticisms of von Braun and twisted inconsequential events into sinister activities. It was virtually ignored west of the Iron Curtain, and von Braun's membership in the SS and involvement in activities at Dora remained secret.[17]

Von Braun also took a beating in Moscow by the government-controlled newspaper *Izvestia*. The *Washington Post* reported *Izvestia*'s charges on August 6, 1964: "Izvestia . . . attacked American rocket expert Wernher von Braun for 'creating inhuman conditions' among workers at an underground Nazi rocket base in Poland [*sic*] that he reportedly headed during World War II. Von Braun, *Izvestia* said, did so 'to carry out the promise given to Hitler of a miracle-weapon.' "[18] The *Izvestia* article had many factual errors: Von Braun did not control working conditions; the rocket base (Peenemuende) and the underground factory (Mittelwerk) were both in Germany.

While von Braun could ignore attacks from behind the Iron Curtain as unfounded, libelous, Cold War rhetoric, he was compelled to address charges made against him by readers of the widely circulated and respected *Paris Match*. The occasion was triggered by von Braun's own self-promotion machine. *Paris Match* had published several complimentary pieces on von Braun and his team, and in the fall of 1965 it sent its writers to the United States to get a better view of the space program. Von Braun favored these reporters and other members of the European press with a test firing of the F-1 rocket engine, five of which would power the first stage of the Saturn V moon rocket. *Paris Match* returned the favor with a flattering report in its October 23, 1965, issue.

The article elicited a flood of letters to the editor from members of a group calling itself the *Amicale des Camps de Dora-Ellrich* (Friends of Deportees of the Dora-Ellrich Camps). They held von Braun partly responsible for the suffering of inmates of the concentration camp where the V-2 was built, and they were offended that the man most responsible for creating the weapon was held in such high regard. By April 1966, the stack of letters was too high to ignore, and *Paris Match* asked von Braun to answer his accusers. Von Braun replied in a letter dated April 26, 1966.

After twenty years passed without being held accountable for his actions on behalf of Nazi Germany, Wernher von Braun was in deep denial: "As much as I understand their bitterness, I am appalled by their false accusations aimed at me." He believed that the victims of Dora and the Mittelwerk were now making him a victim. He went on to explain how the United States government had carefully studied his background before his immigration and how a war-crimes tribunal had investigated the atrocities of Dora, yet neither investigation had found information that reflected adversely on him. He had reason to believe that he was in no way responsible or accountable for the crimes of the Third Reich.

About the crimes at Dora, von Braun wrote in his letter to *Paris Match*, "I felt ashamed that things like this were possible in Germany, even under a war situation where national survival was at stake."

Von Braun could not help but cite his arrest by the Gestapo on a charge that he and his colleagues had sabotaged the V-2 development program as evidence that he was not in league with the operators of Dora. However, von Braun failed to mention that his arrest was part of Himmler's power play to gain control of the V-2 program and that most of his activities at Dora and the Mittelwerk took place after his arrest in March 1944. The editors of *Paris Match* were apparently satisfied with von Braun's answer— or more interested in the personalities and fashions of the day—and let the issue drop. Once again von Braun escaped public scrutiny of his activities in Nazi Germany.[19]

The exploration of space was dangerous, although those involved did not like to talk about it. In the mid-1960s, the dangers were exacerbated by actions of incompetents and stupid blunders. Death claimed some of the key players of space exploration.

One of the first to be lost was Sergei Korolyov, the chief designer of the Soviet Union's space program. Korolyov's health was eroded by his many years in Stalin's gulag, and, reportedly, he had a weak heart. Ironically, the problem that led to his death was hemorrhoids. On January 14, 1966, Korolyov was scheduled to undergo surgery to solve the minor problem, and, because of his stature, he acquired as his surgeon Dr. Boris Petrovskiy, minister of health of the Soviet Union. As Petrovskiy operated, he discovered evidence of colon cancer, and despite his lack of recent surgical practice and insufficient supplies and equipment, he proceeded to remove the tumor. Petrovskiy severed a large blood vessel, and Korolyov bled to death before a competent surgeon was found to correct the damage.

Two days later, *Pravda* announced the death of Sergei Korolyov, omitting, of course, any mention of the medical malpractice that caused his death. Korolyov was cremated, and his ashes were given a place of honor in the Kremlin Wall. Only in death was Korolyov known and honored as the chief designer of the Soviet space program.[20]

While Korolyov was lost, his influence on the Soviet space program continued. The Soyuz launch vehicle that he created (a derivative of the Semyorka design) is to this day a workhorse and critical component of the Russian space program.[21]

The United States also had its share of pointless tragedy. The first manned flight of an Apollo spacecraft was scheduled for February 21, 1967. It would carry a crew of three and be launched into orbit by a Saturn IB. The spacecraft had been placed atop the booster at launch pad 34 at the Kennedy Space Center and was going through its final checks. In the afternoon of January 27, the three astronauts, Roger B. Chaffee, Edward H. White, and Virgil I. (Gus) Grissom, wearing space suits, entered the command module for a full dress rehearsal of the final countdown. The hatch was bolted closed, and the capsule was pressurized to 16.7 pounds per square inch with pure oxygen, as it would be prior to a real launch. Unlike a real launch, however, the rocket stages were not fueled.

As the crew at Kennedy Space Center was checking-out the Apollo capsule, the top executives of the Apollo program were attending their quarterly meeting at NASA headquarters in Washington, D.C. The participants in the meeting were James Webb, NASA administrator; Wernher von Braun; Robert Gilruth, director of the Manned Spacecraft Center at Houston; Kurt Debus, director of Kennedy Space Center; General Samuel Phillips, Apollo program director; George Mueller, NASA administrator for manned space flight; and the chief executives of the prime contractors for the Apollo program. After they finished the work of the first day of a two-day meeting, they went over to the White House for a celebratory reception. That afternoon in the East Room, ambassadors from sixty nations had signed a "space treaty," which forbade the use of outer space for military purposes. Following the reception, the group of Apollo executives left for dinner at the International Club on 19th Street, just a few blocks from the White House. They were accompanied by Vice President Hubert Humphrey and several members of the House and Senate Space and Appropriations Committees. After a productive day and the international dedication of space to peaceful purposes, they had reason to celebrate.

They had not yet sat down to dinner when the telephone calls started coming in, telling the horrible story of what had happened at the Cape. Gilruth, Mueller, Phillips, Debus, and the contractors whose hardware was involved, left to fly to the Cape.[22]

At 6:31 P.M., the astronauts had reported a fire in the capsule. The pure oxygen atmosphere fueled a rapid combustion of plastics and other usually fire resistant materials. The three astronauts lost consciousness within thirty seconds and were dead from asphyxiation due to smoke inhalation before the ground crew could open the hatch.[23]

The Apollo 204 Review Board (the number pertains to the manufacturing designation of the Saturn IB rocket, which had no part in the tragedy) investigated the accident and identified the immediate cause of the fire as a spark from a frayed wire. The Review Board's final report was a 3,000-page critique of the entire Apollo program. It cited numerous deficiencies in design, engineering, manufacturing, and quality control. This intense examination of the Apollo program resulted in revisions in the operation of the program and a comprehensive repair of the deficiencies of the spacecraft. The Apollo tragedy that killed three astronauts also resulted in a severe blow to the self-confidence—bordering on arrogance—of those working in the program and a delay of the first manned Apollo flight for two years.

Gus Grissom, one of the dead astronauts had spoken prophetically of a possible tragedy: "If we die we want people to accept it. We are in a risky business, and we hope that if anything happens to us it will not delay the program. The conquest of space is worth the risk of life."[24]

The Soviet space program took the next turn with death on April 24, 1967. After one day in orbit in the Soyuz-I, Col. Vladimir Komarov died as the spaceship crashed to earth. The parachute lines had tangled during the final descent.[25,26]

Yuri Gagarin, the first man to enter space and orbit the earth, followed Komarov in death in March 1968 while on a routine jet training flight. The accident investigation files attributed the crash to "pilot error." [27] Vladimir Komarov's and Yuri Gagarin's ashes found their final places of honor near those of Sergei Korolyov in the Kremlin Wall.[25,26]

Despite these tragic losses, the goal was still there, the spaceships were still being developed, and the astronauts and cosmonauts were still ready. The race for the moon would go on. The living could rationalize that the dead had perished because of some fatal flaw in themselves: Korolyov chose the wrong surgeon, the three Apollo astronauts and Komarov had willingly entered flawed spacecraft, Gagarin was guilty of "pilot error." The dead were somehow deficient in what Tom Wolfe would later describe—with reference to American pilots—as "The Right Stuff."[28] Of course, those in line to fly Apollo knew that they were the best, that they would triumph where others failed, and that they would walk on the moon.

In 1968, the United States blundered through a series of political crises and public disasters that distracted public attention from its struggling space program. Under the leadership of President Lyndon Johnson, the country had blundered its way into the Vietnam War. In January and February, Vietcong guerrillas pointed out how badly the United States understood their resolve by attacking Saigon and the major cities of South Vietnam in what became known as the Tet offensive. President Johnson was so shaken by the course of his unpopular war that, at the end of March, he announced that he would not be a candidate for reelection that year. Less than a week later, on April 4, civil-rights leader Martin Luther King Jr. was assassinated. On June 5, after winning the California primary, Robert Kennedy, the leading Democratic candidate for the presidency, was shot in Los Angeles; he died the following day. When November rolled around, with the Democrats in disarray, Richard M. Nixon defeated Hubert Humphrey and won the presidential election.[29]

By the end of 1968, the Apollo program was back on course, driven by the commitment of the hundreds of thousands of people who worked on it and the fear that the Soviet Union would beat the United States to the moon. On October 11, a Saturn IB carried *Apollo 7*, the first manned flight of the Apollo program, into earth orbit. *Apollo 8* was scheduled for liftoff on December 21. It would be boosted by a Saturn V onto a course to the moon. There it would spend a short time in lunar orbit before returning to earth.[30]

With the final days of his presidency slipping away, Lyndon Johnson invited the astronauts of *Apollo 8* and their wives to the White House for dinner. Johnson also invited Charles Lindbergh and Wernher von Braun and their wives. The legends of the rapidly fading past, Lindbergh, von Braun, and Johnson, met the heroes of the future, astronauts Frank Borman, Jim Lovell, and William (Bill) Anders about two weeks before their journey. Frank Borman, who would command the flight recalled that at the dinner, President Johnson was preoccupied not with the flight of *Apollo 8*, but with the bloodletting of the war in Vietnam and the discord it had caused.[31]

Johnson's preoccupation that evening mirrored that of the American people. Wernher von Braun's most enthusiastic constituency, American youth, lost much of its previous interest in space exploration. In a cultural wave that surged through the decade parallel with manned exploration of space and deepening entanglement in the Vietnam War, the young poured their passion into what they quaintly called the counterculture. Better-educated young men had no interest in trading college for a tour of duty in Vietnam. Young men and women heeded the call to turn on, tune in, and drop out. And while 1968 was not a good year for much else, it was, while it lasted, a good year for sex, drugs, and rock and roll.

16

Men on the Moon

Wernher [von Braun] ... postulated, predicted, advertised, conned, pulled and finally pushed to make us first on the moon.
Neil Armstrong, Buzz Aldrin, Mike Collins[1]

History was the past, made of pieces laboriously fit together by scholars in the hope of producing a comprehensible picture. It existed mostly in books and scholarly journals enshrined in dusty archives. Those who had observed historical events commonly saw only a small part of the picture, even if they were participants. Consider the fall of Troy, the burning of Rome, Columbus's discovery of the New World, or the explosion of the atom bomb at Hiroshima. Days or decades passed before witnesses told their stories, facts became known, and their significance was understood. Mankind's first journey to another celestial body, the flight of *Apollo 11* to the moon, changed all that. Its occurrence had been decreed by John Kennedy eight years before; all aspects had been planned to the second and rehearsed to a knowing certainty by all participants. Furthermore, the United States had made it a matter of national honor to accomplish the journey—as it had done for its entire space program—to show the technological and moral superiority of democracy over the Communist system of the Soviet Union.

For the first time, with the flight of *Apollo 11*, history became a spectator sport. Those so inclined, and about a million were, showed up at Cape Canaveral and experienced it with their own senses. Those who weren't at the Cape watched it on television, sitting on overstuffed sofas, sipping beers, just as if they were watching the World Series or the Superbowl.

Once the Apollo program recovered from the tragedy of *Apollo 1*, it moved quickly, if not smoothly, in preparation for the manned lunar landing mission. The launch of the *Apollo 7* spacecraft into earth orbit on a Saturn IB rocket on October 11, 1968, was the first manned flight of the Apollo spacecraft. On December 21, a Saturn V boosted *Apollo 8* and its three crewmen on the first manned voyage around the moon. On March 3, 1969, a Saturn V carried command and lunar landing modules into earth orbit for a rehearsal of docking in space. Then on May 18, a Saturn V lifted the *Apollo 10* spacecraft again toward the moon; this time two crew members guided the lunar module to within nine miles of the lunar surface.

If the landing on the moon were dependent on the reliability of the rockets alone—the rockets designed and built by Wernher von Braun, his team, and the United States aerospace industry—it was virtually guaranteed to be a success. Before *Apollo 11*, the Saturn I rocket in several versions had made fifteen successful flights, and the Saturn V had performed with near-perfection on five flights.[2]

Spectators began to arrive on the Atlantic coast of Florida at least ten days before the scheduled launch of the Saturn V/*Apollo 11*. The press started arriving soon after. By Sunday, July 13, three days before liftoff, most of the press corps was on the scene. With three days to the launch, the parties began with lots of daiquiris and dancing.[3]

Members of the press managed to make time to attend press conferences with the astronauts, and on Monday, July 14, several hundred attended a conference with ranking officials of NASA. The press met a panel composed of George Mueller, NASA associate administrator for Manned Space Flight, Robert R. Gilruth, director of the Manned Spacecraft Center at Houston, Kurt Debus, director of the Kennedy Space Center, and Wernher von Braun. It seemed that half of the questions were posed to von Braun, a testament to the fact that the fifty-seven-year-old engineer and bureaucrat was the most recognizable participant in the space program who was not an astronaut.[4]

While most reporters paid attention to the questions and answers, one correspondent, novelist and social observer Norman Mailer, paid attention to the man. Mailer had the energy and ego to match von Braun's. Of Wernher von Braun and his presence at the press conference, Mailer wrote:

Since he [von Braun] had in contrast to his delivery, a big burly squared-off bulk of a body which gave hint of the methodical ruthlessness of more than one Russian bureaucrat, von Braun's relatively small voice, darting eyes, and semaphoric presentation of lip made it obvious he was a man of opposites. He revealed a confusing aura of strength and vulnerability, of calm and agitation, of cruelty and concern, phlegm and sensitivity, which would have given fine play to the talents of so virtuoso an actor as Mr. Rod Steiger. Von Braun had in fact something of Steiger's soft voice, the play of force and weakness which speaks of consecration and vanity, dedication and indulgence, steel and fat.[5]

Who knows whether Mailer's evaluation of von Braun was accurate. Still, it was refreshing because it tried to probe beneath the obvious and reach the character of the man. Over the decades, most Americans had been willing to take von Braun at face value, and it was a portrait created by von Braun himself.

During the question and answer period, a reporter from East Berlin asked von Braun a question in German. After an uncomfortable moment, von Braun translated the question into English, then gave a long and detailed answer in English. For the benefit of the reporter from East Berlin, he also gave his answer in German. Then, having bored most of the press in attendance, he added, "I must warn the hundred and thirty-four Japanese correspondents here at Cape Kennedy that I cannot do the same in Japanese."[5]

During the press conference, von Braun was asked to evaluate the significance of putting a man on the moon. Ever the pitchman, he answered, "I think it is equal in importance to that moment when aquatic life came crawling up on the land." Whether true or not, von Braun's observation was featured in press reporting.[6]

The three astronauts who were to make the trip were in relaxed seclusion. Soon after they returned, they took their turn at writing the history they had created, and they did a better job than most in characterizing the mania that boiled around them:

In Cocoa Beach and the neighboring community, Cape Canaveral, the tension was becoming indistinguishable from physical pain. . . . There is something unreal, something unearthly, about the place; it is as if all its creature comfort facilities are maintained on standby. . . . Cocoa Beach is a rather dull place between "shots," and the motels are almost always glad to welcome the odd tourist who shows up without an advance reservation. But "shot time" is a wild time in Cocoa Beach. A special kind of hysteria seems to seize the place; it is as if a whole mass of people had gone on LSD for a predictable number of days. People drink abnormal amounts of hard liquor, eat irregularly or not at all and hardly ever go to bed. . . . Private parties

proliferate to lavish proportions; most of them are sponsored by contractors or prime news media in the interest of "image" public relations, but some are given by private citizens of means. At shot time
Cocoa Beach exerts its own gravitational pull.[7]

The night before the shot, every motel within driving distance was the
site of a countdown party.[8] The big party, the one that was attended by
the rich, the famous, and the powerful, was given at the Royal Oak Country Club in Titusville, twenty miles north of Cocoa Beach. The party was
paid for by *Life* magazine,[9] which had sponsored a junket for corporate
presidents and ranking executives (who presumably controlled large advertising budgets), to go first to Houston to meet with the astronauts, and
then to the Cape to witness the launch of the Saturn V/*Apollo 11.*[10]

The honored speaker at *Life*'s pre-launch banquet was Wernher von
Braun. *Life* gave him what might have been one of his last opportunities
to collect the glory, cash in on speaking fees, and pitch for the future of
the space program—if there was to be one after the moon. In a few weeks,
the headliners on the speaking circuit would be the three men who would
sit atop von Braun's Saturn V the following morning—if, of course, they
landed on the moon and returned home. Von Braun, with his wife Maria,
arrived by helicopter, and he quickly moved into the hexagonal-shaped,
walnut-paneled banquet room.

During the round of hand-shaking that preceded the banquet, Norman
Mailer, who had been contracted by *Life* magazine to write about the first
expedition to the moon, met von Braun. Von Braun, still hustling funds
for the exploration of space said to Mailer, "You must help us give a *shove*
to the program. . . . Yes." Von Braun tried to buy Mailer's good will with
a smile, "we are in trouble. You must help us."

"Who are you kidding?" Mailer said. The moon landing would be von
Braun's triumph. Congress couldn't help but pour in the cash for the next
project. "You're going to get everything you want." Mailer was apparently
better at taking the measure of individuals than political bodies.

This was too much for von Braun, who uncomfortably moved on to the
next guest, who, presumably, would have better manners than to speak his
mind.[11]

Soon, the magazine's guests at the pre-launch party, the captains of
American industry, moved on from cocktails to roast beef, to ice cream,
and then to the featured speaker. The publisher of *Life*—the man who
would sign the check for the banquet—took the podium to begin the in-
troductions. First, he recognized not Wernher von Braun, but his early
teacher, Hermann Oberth. He described Oberth accurately as one of the
fathers—along with the Russian Tsiolkovsky and the American Goddard—
of space exploration. At seventy-five, Oberth's hair had gone completely

white, yet he still had hard-edged features and a proud demeanor. He was a living legend, though few remembered him.

When the publisher introduced von Braun, he went into the boring, yet potentially embarrassing details of von Braun's service to the country of his birth, Germany, but rescued the moment by reminding everyone that von Braun had become a United States citizen in 1955. As the publisher finished and von Braun moved to the fore, the audience gave him a standing ovation. He may have been a Nazi, but he was America's ex-Nazi.

Von Braun began by thanking the corporate executives and sharing the credit with them for the great event that was to come. And lest anyone forget, he reminded them that "It is an American triumph." When von Braun got into the core of his discourse, he seemed to write off the landing of a man on the moon as an insignificant act. What was the moon after all, but a lump of gray rock, dead and of fleeting interest. The significance of the flight of Saturn V/*Apollo 11* was that it would be mankind's first step into new worlds of existence and thought. He said, "What we are seeking in tomorrow's trip is indeed the future on earth. We are expanding the mind of man. We are extending this God-given brain and these God-given hands to their outermost limits and in so doing all mankind will benefit. All mankind will reap the harvest. . . . What we will have attained when Neil Armstrong steps down upon the moon is a completely new step in the evolution of man."

"A new step in the evolution of man." Hyperbole. Lousy science. Too vague for theology. The cosmic concept was compressed into an easily remembered phrase.

When the questions began, von Braun had his wish list ready. The future of the space program would be reusable transport rockets, manned laboratories in orbit around earth, massive nuclear rockets, and expeditions to Mars. The trip to the moon was not the end of the program, but the beginning of a new age of discovery. All he needed was the financial backing. Perhaps his rich and powerful audience could help. After answering a number of polite questions, Wernher von Braun took his leave. He was carried away in his helicopter to greet senators and representatives and continue his lobbying for funds.[12]

Von Braun returned to his room at Cocoa Beach. He was restless, and spent an hour reviewing the launch schedule for the Saturn V/*Apollo 11*. He telephoned his old friend and colleague Kurt Debus to wish him good luck on the following day's launch and to ask about a few minor details. Then, at last, he slipped into bed and closed his eyes.[13]

To be fair, the million or so people who came to the Cape were not there primarily for the parties, but to share in the great adventure that could be experienced first-hand by only three men. Just as the sun rose, Neil Armstrong, Mike Collins, and Buzz Aldrin entered the command module of the Saturn V stack. (From the ground up, the Saturn V stack was composed of the S-IC first stage, the S-II second stage, the S-IVB third stage, a three-foot-high instrument unit, the lunar module, the conical command and service module, and the emergency escape tower). They had rehearsed their parts since they had been assigned to the flight in January 1969. As they sat near the apex of the 363-foot-high tower, everything was the same as in their exercises, except that they were now strapped to a potentially explosive reservoir containing more than six million pounds of liquid oxygen (LOX), kerosene, liquid hydrogen, and self-igniting hypergolic fuels.[14] And, of course, the whole world was watching.

Everyone was acting as if there was little or no risk in the adventure. This had in fact been planned. In the early stages of the Apollo program, NASA defined the probability of returning the astronauts safely back to earth as 0.999, that is, 999 times out of one thousand. This probability was expressed verbally as "three nines." Likewise, the probability of completing the assigned mission of landing on the moon and returning safely was set at 0.99, 99 times out of one hundred or "two nines." In an anecdote that made the rounds at NASA, a group from headquarters went to Huntsville to ask von Braun and his team about the reliability of the Saturn V. Von Braun turned to his senior staff and asked, "Is there any reason why it won't work?" They answered in turn, "Nein." "Nein." "Nein." "Nein." Von Braun then said to his visitors from headquarters, "Gentlemen, I have a reliability of four nines."[15] On the morning of July 16, everyone accepted the reliability of the rocket. Nobody dared consider the possibility of failure.

The audience was immense. In round numbers, one newsmagazine estimated that over one million people were stretched out along Florida's eastern coast within view of the Saturn V/Apollo 11.[16] Most came in cars, trucks, and campers. Thousands were aboard approximately 3,000 boats anchored in the Indian and Banana Rivers near the Cape. Seventeen hundred and eighty-two journalists were there to make sure that history was properly recorded. Some 6,000 guests were there by NASA's invitation. These included 205 congressmen, 30 senators, 19 governors, 50 mayors, and 69 ambassadors. Lyndon and Ladybird Johnson were there as representatives of President Nixon.[3] Also present were Vice President Spiro Agnew,[16] Charles Lindbergh,[3] and Hermann Oberth, whose theories and plans had started the chain of events nearly a half-century earlier.[17] Those who did not make it to the Cape watched on television or listened to the radio. ABC-TV, which managed the pool coverage, estimated that 528 million people followed the launch.[18]

At 4:00 A.M., Wernher von Braun arrived at Launch Control Center at the Kennedy Space Center. He took an elevator to the huge control room where over fifty men sat at banks of instrument panels guiding the spacecraft through its countdown. Kurt Debus told him that the countdown was proceeding smoothly. Von Braun then went to the adjacent glass-enclosed room where VIPs could feel part of the action without getting in the way.[13] He was joined there by Dr. George Mueller, NASA associate administrator for Manned Space Flight, and Lt. Gen. Samuel C. Phillips, director of the Apollo Program, and a few others. Through blast-proof windows that stretched from near the floor to the tall ceiling, they used binoculars to watch the lift-off from launchpad 39-A, four miles away.[19]

"T minus sixty seconds and counting."[20] The voice of Apollo-Saturn Launch Control is on every loudspeaker at the Cape and on most radios and televisions around the earth.

Von Braun's dream is becoming a reality, and is now in the hands of others.

"T minus fifteen seconds, guidance is internal."

"So," von Braun says softly. Then he does the only thing left for him to do. He silently prays, *Our Father, who art in heaven.*[19]

"Twelve, eleven, ten, nine, ignition sequence starts."

Hallowed be thy name. Thy kingdom come. Thy will be done on earth, as it is in heaven.

"Six, five, four, three, two, one, zero, all engines running."

Give us this day our daily bread. And forgive us our debts, as we forgive our debtors.

"LIFTOFF! We have a liftoff, thirty-two minutes past the hour."

For Thine is the kingdom and the power, and the glory, for ever.

"Liftoff of Apollo 11."

Amen.[21]

Three miles away, the closest spectators, the press, watch the rocket clear the top of the launchpad tower. Only then, fifteen seconds after the Saturn V's huge engines ignited and six full seconds after they achieved enough thrust to liftoff from the pad trailing a cataract of flame, does the press corps hear the crackling of the engines. The sound builds to a near-deafening staccato of explosions, and the earth begins to shake from the violent power and will not stop.[22]

The rocket climbs, leans out over the Atlantic, and keeps ascending into the sky. Soon, those on the ground can no longer see the spaceship, only the pounding flame that pushes it higher and faster.[22] Less than three minutes into the flight, the first stage has burned all of its fuel and drops off into the ocean. The second stage ignites and continues the ascent. Three minutes after liftoff, the emergency escape tower shoots free. The second stage continues to burn until nine minutes and eleven seconds after liftoff, then it, too, drops away. The single engine of the third stage burns for two and one-half minutes, then shuts off, leaving the spacecraft with its three passengers floating in earth orbit.[23]

Two hours and forty-five minutes after liftoff, *Apollo 11* is well into its second orbit of the earth. The astronauts restart the J-2 engine of the third stage, the S-IVB. It burns silently, but pushes the astronauts into their couches with a force of 1.5 Gs. Six minutes later, the third stage engine stops, and *Apollo 11* is traveling at 35,570 feet per second (about 25,000 miles per hour) on a course that will intersect with the moon.[24]

At this point, Mike Collins pilots the command module free of the remainder of the spaceship, rotates it 180 degrees, and docks its conical point to the top of the lunar module. The S-IVB third stage, still attached to the bottom side of the lunar module, will be separated only after completion of a long checklist.[24]

The *Apollo 11* spacecraft coasts toward the moon, losing speed because of earth's gravity pulling it back, then increasing in speed because of the greater gravitational effect of the moon. Seventy-six hours after liftoff, *Apollo 11* curves behind the moon, guided by its momentum and the moon's gravity. Mike Collins pushes the control button, and the rocket engine of the command and service module fires for six minutes. *Apollo 11* achieves lunar orbit.[25]

Scarcely noticed by the West during the flight of *Apollo 11*, the Soviet Union is also racing to the moon. On July 13, three days before the departure of *Apollo 11*, the Soviet Union launches *Luna 15*, an unmanned probe designed to land on the moon and return to earth with a small sample of lunar soil. It arrives at the moon and parks in lunar orbit two days before the arrival of *Apollo 11*. The orbit is such that it will not interfere with the manned United States mission. *Luna 15* maneuvers for a soft landing on the Mare Crisium, the Sea of Crises. It crashes, destroying itself and the hopes of the Soviet Union to beat its American rivals in the achievement of one scientific goal, the retrieval of a sample of the moon for scientific study.[26,27] This challenge is now in the hands of Armstrong, Aldrin, and Collins.

The flight of *Apollo 11* is guided from the Manned Spacecraft Center in Houston. Looking over the shoulders of the flight controllers in Mission Control are NASA Administrator Thomas O. Paine and several members of his staff, a large number of astronauts, past and present, and, of course, Wernher von Braun. Everybody seems to be sweating, and not simply because of the typical steamy Houston summer weather. All their planning is about to reach its conclusion, either in triumph or disaster.

Neil Armstrong and Buzz Aldrin enter the lunar module and guide it down to a lower orbit, the stage for the final descent. Descending the last eighteen miles to the surface of the moon take about a dozen minutes. With Houston's approval, the lunar module crew fires its rockets, and begins their descent. The lunar module's onboard computer is programmed to control the retro-rocket and guide the module to the surface of the moon. Partway through the descent, the computer flashes an alarm code. The descent continues while the staff at Houston try to figure out what is going on. They decide that the computer is overloaded with data; not a big deal—unless, of course, it is the first descent ever to the moon's surface and the onboard computer is flashing an alarm code. A few hundred feet above the surface, Armstrong sees that the computer is guiding them into a boulder field. He takes over manual control and begins searching for a clear area in which to set down the lunar module. With only an estimated twenty seconds of fuel left in the lunar module's tanks, Armstrong says, "Houston, Tranquillity Base here. The Eagle has landed."

The time, 4:18 P.M. eastern daylight time, 3:18 P.M. in Houston.

Back on earth, they begin to breathe again.[28,29]

On July 21, after 21 hours and 38 minutes on the moon, 2 hours and 21 minutes actually walking on the lunar surface, the astronauts aboard the Eagle lunar module blast off on their return to earth. Armstrong and Aldrin had planted the flag, left behind some experiments and mementos, and collected about forty pounds of moon rocks. Their rendezvous with the command module goes according to plan, and they steer a course back to earth.[30,31]

On the morning of Tuesday, July 22, the Soviet Union announced ambiguously that *Luna 15* had "reached the moon's surface" and that its mission had "ended."[32]

In the early morning of July 24, the *Apollo 11* command module splashed down in the Pacific Ocean some 950 miles southwest of Hawaii, near the spot where the aircraft carrier *Hornet* was waiting to retrieve it and its three crewmen. The president of the United States was there to greet them.[33]

The three astronauts had invited President Nixon to join them for dinner at the Kennedy Space Center on the evening before their flight, but he began to show symptoms of a cold and graciously and prudently declined their invitation. If any of the three astronauts had even sneezed on their trip to the moon, it was clear who would be blamed. But Nixon, like almost everyone else on earth, was caught up in the excitement of the moon landing, and he went almost half-way around the world to greet the astronauts when they boarded the *Hornet*.

Some at NASA were concerned about the very remote possibility that the *Apollo 11* astronauts might return from the moon carrying a foreign microbe that might cause a lunar plague on earth. When the astronauts boarded the *Hornet*, they were escorted directly to a customized aluminum trailer parked on the hangar deck, where they would begin their isolation for the next eighteen days. They had about forty minutes to be sampled for microbes, native to earth and otherwise, and take showers before they faced the president through a window in the quarantine trailer. He was effusive in his praise: "Neil, Buzz and Mike, I want you to know that I think I'm the luckiest man in the world, and I say this not only because I have the honor to be President of the United States, but particularly because I have the privilege of speaking for so many in welcoming you back to earth."[34]

Nixon invited the astronauts and their wives to a state dinner after they got out of quarantine. In the excitement of the moment, and in the glare of the lights for the television cameras, he was reluctant to say goodbye. He also did most of the talking.

Normally a cool observer of events, Nixon drifted into hyperbole. "This is the greatest week in the history of the world since the Creation, because as a result of what happened, in this week, the world is bigger infinitely, and also . . . the world's never been closer together before."[34]

The success of the *Apollo 11* flight brought about several ironic transitions. Von Braun and the engineers were replaced by three astronauts as the true explorers of space, and Richard Nixon replaced John Kennedy as its patron. Regrettably, the glory and excitement did not last.

After their successful return to earth, the three *Apollo 11* astronauts became national treasures. They were cherished, protected, and used to promote a progressive image for NASA and the United States. The government would never again willingly allow them to fly anything riskier than a commercial airliner. Faced with the end of their careers as pilots, they all eventually left the space program. Richard Nixon did not carry his excitement through to financial support of subsequent space projects.

Public interest in the manned moon-landing program waned after the success of *Apollo 11*. John Kennedy's impressive goal had been achieved. Men walked on the moon, collected lunar rocks, and came back home safely. In the two and a half years that followed, six more Apollo missions flew to the moon, and with the exception of the aborted flight of *Apollo 13*, the flights were carried out, with occasional minor deviations, according to plan. The flight of *Apollo 11* was the grand media event of the century; the Apollo missions that followed played on television like reruns.

When the *Apollo 11* command module returned to earth, Wernher von Braun was back home in Huntsville. The news of the success—Huntsville's success, since the Saturn V rocket had been conceived there—triggered a massive celebration, and the townspeople carried von Braun through the streets on their shoulders.[35] Two days later, the staff of the Marshall Space Flight Center celebrated the lunar landing with an afternoon picnic at which von Braun shared the glory of success with the staff of Marshall.[36] That evening he shared the glory with Huntsville's civic leaders at an evening banquet.[37]

President Nixon planned a White House celebration for August 7, and intended to invite the key Apollo team members including von Braun. Before sending out the invitations, however, Nixon's assistant for domestic affairs, John Ehrlichman, asked the FBI to conduct a security check on von Braun. Ehrlichman received a letter from FBI director J. Edgar Hoover that stated, "Although these investigations were generally favorable and indicated that Dr. von Braun was anticommunist, information was developed that he received an honorary SS Commission as a Lieutenant and had been a member of the National Socialist Party (Nazi) in 1939."[38] After twenty-four years working for the government, after fourteen years as a citizen, after he led the development of the rockets that led to the nation's triumph in space, they still did not trust him.

President Nixon's celebration ultimately took place on Wednesday, August 13, 1969, after the *Apollo 11* astronauts emerged from their lunar-expedition quarantine. The astronauts began their day with a parade down New York's Broadway, continued with a motorcade and rally in Chicago, that culminated with a formal dinner hosted by the president in Los Angeles.[39] The dinner at the Century Plaza Hotel was televised so that all Americans could join President Nixon and his 1,440 guests, among whom were Dr. and Mrs. Wernher von Braun of Huntsville, Alabama.[40,41]

17

Shuttle, Space Station, and the Decline of NASA

> Other journeys must follow [the first landing on the moon]. We must use the Saturn rockets, the Apollo spacecraft and the launch facilities built up in Project Apollo over and over again to gain the fullest return on our investment. To make a one night stand on the moon, and go there no more, would be as senseless as building a locomotive and a trans-continental railroad, and then making one trip from New York to Los Angeles.
>
> Wernher von Braun[1]

The year 1969 should have been a happy one for Wernher von Braun since it was, after all, the year in which his dream of space exploration reached its climax with a manned landing on the moon. The future of space travel was finally opening up, but the ghosts of his past were in pursuit.

His troubles began in early 1969, when West German prosecutors investigating allegations of war crimes committed at Dora concentration camp asked to interview him. Although von Braun may not have received explicit instructions to submit to an interview, as an employee of the federal government, he could not easily refuse an allied government agency. The information he gave to the West German prosecutors on February 7 had the same see-no-evil, hear-no-evil, speak-no-evil character[2] as did his deposition to the Dora-Nordhausen War Crimes Trials in 1947.[3] "I never saw a dead man, nor maltreatment nor killing," he told them. Regarding reports in Peenemuende files of sabotage by prisoners at the Mittelwerk, he said, "I cannot remember anything about it."[2] This inquiry may have been disturbing, but it was nothing new. He had answered similar questions in his deposition in 1947 and again in accusations in the French press in 1965 and 1966, and escaped serious harm by protesting his ignorance and condemning the crimes. Still, having to deal with the subject must have taken some of the luster off the upcoming launch of *Apollo 11*.

About the time of the *Apollo 11* launch, retired Maj. Gen. Julius Klein asked von Braun about an accusation made against him by muckraking columnist Drew Pearson. About two decades earlier, Pearson alleged that von Braun had been a member of the SS, though few paid any attention or remembered the accusation. Von Braun, however, appeared to be shaken by Major General Klein's private inquiry. It was the first time in decades that the issue had been raised by anybody. "It's true that I was a member of Hitler's SS elite. The columnist was correct," he wrote. "I would appreciate it if you would keep the information to yourself as any publicity would harm my work with NASA."[2] Klein, apparently swayed by von Braun's plea, did not pursue the issue.

After the successful return to earth of the *Apollo 11* astronauts, President Nixon planned a dinner at the White House to celebrate the triumph and wanted to include Wernher von Braun. Despite his years of service to the federal government and the security clearances he received because of his service, Nixon's staff required a security check from the FBI before inviting von Braun (see Chapter 16).[4]

Von Braun may have been unaware of the White House security check, but he could not have missed the protest directed at him six weeks later in Wilmington, Delaware. The Golden Slipper Square Club, a predominantly Jewish organization, invited von Braun and NASA officials to a banquet on September 16. They were greeted by a group calling themselves the Wilmington Survivors, a group of Jewish concentration camp survivors who refused to forget. About twenty five pickets paraded outside the club carrying signs referring to von Braun's Nazi background and walked over a Nazi flag. Having made their point, the group left peacefully.[5]

In late January 1969, Cornelius Ryan, who edited the series of *Collier's* articles that placed von Braun's plans for space flight before the public seventeen years earlier, wrote to his old friend and colleague. *Collier's* was long since a casualty of the competiveness of magazine publishing, but Ryan was now with *Reader's Digest*. The Apollo missions were likely to take astronauts to the moon and return them safely to earth before the end of the year, but what would follow Apollo? Ryan asked von Braun to write an article describing his concept of the future of the United States space program. Von Braun could not pass up the opportunity to once again be a prophet.

Von Braun wrote his article and revised it several times by early October 1969, after the flight of *Apollo 11*. He defined a broad and ambitious agenda that included completion of the entire Apollo program for the exploration of the moon, the use of unmanned robot probes to explore the planets, communication and earth observation satellites, a reusable space shuttle, earth-orbiting space stations, and a manned expedition to Mars. Von Braun wisely gave no timetable or cost estimates for any of these projects. He concluded his article by drawing a parallel between NASA and Henry the Navigator (1394–1460), a Portuguese prince. Henry's claim to immortality was his establishment of an observatory and the first school of navigators in Europe; he was also responsible for improvements in shipbuilding and the exploration of the west coast of Africa. After his death, exploration continued with the discoveries of the Cape of Good Hope, the Indian Ocean, and a new trade route to the East.

Von Braun ended the draft of his article as follows:

Henry the Navigator would have been hard put had he been requested to justify his actions on a rational basis, or to predict the pay-off or cost effectiveness of his program of exploration. He committed an act of faith and the world became richer and more beautiful as a result of his program. Exploration of space is the challenge of our day. If we continue to put our faith in it and pursue it, it will reward us handsomely.[6]

It is likely that von Braun privately cast himself as a modern day Henry, although he publicly credited NASA with the vision.

At the end of January 1970, Ryan had a final draft of the manuscript, which he submitted to his editor for approval. After reading it, the editor in chief decided not to print the article, sensing that it did not fit the mood or interests of his readers. Ryan responded with a note that revealed his own misgivings. He wrote, "Seriously, cynical as that might be, the truth is that the tone of this country right now, in my honest opinion, is not very conducive to large expenditures on the space program. Von Braun, in spite of all his protestations, would like to keep on spending like a drunken sailor because, after all, space and its exploration is his life-time dedication." Ryan finished the note to his boss by saying that he would take von Braun to one of New York's best restaurants, "gaze into his deep, blue Teutonic eyes, put a drink in his fist," and, as gently as possible, let him know that *Reader's Digest* would not be publishing his new plan for space.[6]

On February 13, 1969, one of Richard Nixon's first official acts after taking office as president, was to establish the Space Task Force to define long-range goals and to recommend a coherent, long-range space program for the country. Following tradition—short as it was—he put the vice president, Spiro Agnew, in charge of a group that counted as members the new NASA administrator, Thomas O. Paine, presidential science advisor Lee A. DuBridge, and Air Force Secretary Robert C. Seamans.[6,7] One might not have expected much of the task force since, as Thomas O. Paine said of Agnew, "His principal interest [in the space program] was in playing golf with the astronauts."[8] Nevertheless, Agnew drove the task force to its goal. The day *Apollo 11* blasted off for the moon, Agnew, speaking for the task force, defined as a new national goal a manned landing on Mars before the year 2,000.[7] Agnew's proposal echoed John Kennedy's call for a moon landing by the end of the decade. In September 1969, the Space Task Force submitted three options for the future of the United States space program:

- An $8–$10 billion per year program involving a manned Mars expedition, a space station in lunar orbit and a 50-person Earth-orbiting station serviced by a reusable ferry, or space shuttle.

- An intermediate program, costing less than $8 billion annually, that would include the Mars mission.

- A relatively modest $4–$5.7 billion a year program that would embrace an Earth-orbiting space station and the space shuttle as its link to Earth.[9]

To put these projected expenditures in perspective, the early Mercury program cost $392 million; the subsequent Gemini program cost $1.6 billion; and the entire Apollo program was budgeted at $24 billion.[10] The most ambitious option, the Mars-satellite-shuttle program, would cost $8–$10 billion per year, only moderately more than the $6.8 billion NASA and other agencies spent on space programs in 1964, at the height of Apollo activity.[11] (Critics pointed out, though, that Apollo cost over $120 for every one of the 200 million men, women, and children of the United States, even if the spending was spread over a decade.)[12]

All of this should have made Wernher von Braun very happy: The Space Task Force recommended that the United States follow the program for the exploration of space he had originally outlined in detail in the *Collier's* articles. With the announcement of the Space Task Force's recommendations, the future of space exploration began to brighten. The program was presented to the president by his own vice president. What could go wrong?

The Alabama Space and Rocket Center, later renamed the U.S. Space and Rocket Center, was opened to the public in 1970,[13] due in large part to the advocacy of Wernher von Braun and his personal assistance in getting exhibits to put on display. Smaller rockets already at the Marshall Space Flight Center and the Redstone Arsenal were simply transported to the adjoining outdoor site that became the Space and Rocket Center. Von Braun wanted a complete Saturn V/Apollo spaceship as the centerpiece of these exhibits. Parts and stages were scattered across the country, and von Braun contrived to bring them together by making them a requirement of a NASA mission. The components, von Braun alleged, were needed to train crews in transporting them over land. The training route covered a two-mile distance from Marshall to the Space and Rocket Center. Once delivered, the training program was completed, and the Space and Rocket Center generously agreed to store the Saturn V/Apollo components on site at no charge to NASA.[14]

Since it opened in 1970, the Space and Rocket Center added Space Camp, a $4.5 million hands-on activity that simulates experiences in space for students, teachers, and adventurous adults. Its facilities include a centrifuge that subjects forty-six passengers to triple the pull of gravity while circling under a planetarium dome, and Space Shot, which fires twelve passengers in a missile up a 180-foot tower at four times the pull of gravity. Understandably, the U.S. Space and Rocket Center claims to be the biggest tourist attraction in Alabama.[13,15] The experiences of visiting the Center and participating in Space Camp appear to owe a lot to the creative concepts Wernher von Braun and Walt Disney pioneered at Disneyland in the mid-1950s.

It is no small irony that Walt Disney created his second entertainment park less than sixty miles from the Kennedy Space Center in Florida, the launch site of all manned space flights made by the United States' space program. When it opened in October 1971,[16] Walt Disney World featured a Tomorrowland section. Since, at that time, rocket flights to the moon from Kennedy Space Center were almost routine, Disney wisely leaped into the future with a "Mission to Mars" attraction. Three years later, "Mission to Mars" replaced "Rocket to the Moon" at the original Disneyland in Anaheim, California.[17] Walt Disney's Tomorrowland became, like his friend Wernher von Braun's Space and Rocket Center, a museum of the imagination.

By the time *Apollo 11* lifted off for the moon in 1969, it was clear that Marshall Space Flight Center had completed its mission, and had not identified any significant projects for the future. NASA administrator Thomas O. Paine realized this and saw that Marshall's director, Wernher von Braun, could either languish in Huntsville or help him chart NASA's future course at its headquarters in Washington, D.C.

Paine's predecessor, James Webb, had consciously tried to minimize von Braun's visibility in Washington. He was concerned that individuals and groups that remembered the suffering of World War II would object in the strongest and most embarrassing way to having an ex-Nazi in a policy-making position at NASA. Specifically, he was concerned about the "Jewish lobby."[18] Paine did not share this concern. He said:

> I think most people felt that he had a damned unfortunate past and nobody liked a Nazi . . . but he had kind of paid his dues and that he really helped us get to the moon in developing the Saturn V and showed himself to be a worthy citizen of this country, and that while we won't exactly forgive and forget, politeness dictates, at least, we won't get into a disgraceful knock down and drag out.[18]

Paine obviously knew only the public version of von Braun's past; he knew nothing of von Braun's membership in the SS or his complicity in the use of slave labor at the Mittelwerk in the construction of V-2 missiles.

Paine recalled that von Braun was at first ambivalent about moving to Washington, but he realized that he had an ally in Wernher von Braun's wife, Maria. She had been stuck with their children in the desert of west Texas, then in northern Alabama. She was ready for a move to the cosmopolitan capital for the sake of her children and her own enjoyment.[19]

Wernher von Braun, in his need for public visibility and his desire to see the American space program continue, and with the support of his wife, accepted Paine's invitation to move to NASA headquarters in Washington. He showed up for work as associate administrator for Planning Future Programs in February 1970.[20] He no longer had a staff of thousands with a budget to match, but he had a mission to sell to the American people and to the Nixon administration the space program he had described in *Collier's* nearly two decades earlier: the reusable launch vehicles, the orbiting space stations, and the mission to Mars.

In March 1970, President Nixon made public his opinion of the Space Task Force recommendations of the previous September. He chose to support the least ambitious option for the future of the space program, the development of a space transportation system, or space shuttle, but he deferred construction of the space station pending development of the space shuttle.[9] Nixon's announcement was an embarrassment to Vice President Agnew, who chaired the Space Task Force, and it was a great disappointment to the management and employees of NASA. NASA administrator Thomas O. Paine saw that his leadership was not effective, and announced his resignation on July 28, 1970.[21,22] With the man who brought him to NASA headquarters five months earlier now gone, Wernher von Braun was left without his most influential ally. His future at NASA became vague and unpromising.

In the fall of 1970, the White House, in the person of deputy assistant to the president, Alexander P. Butterfield, asked for information on von

Braun from the FBI. The request apparently followed up on an allegation made against von Braun that the White House had received, the subject of which is still classified.[23]

On October 5, 1970, senior FBI staff authorized its investigators to interview von Braun. The internal letter of authorization contains the following, which indicated that FBI Director Hoover and his second in command and constant companion Clyde Tolson had made it known that they were not among Wernher von Braun's admirers: "The Bureau conducted applicant investigations on Von Braun in 1948 and 1961. . . . In 1970 it was decided not to invite Von Braun to speak at FBI Communion Breakfast. In this regard, Mr. Tolson [assistant director of the FBI] commented 'He is a phony,' and the Director [J. Edgar Hoover] concurred. Despite this weakness in his character, our investigations have not indicated any disloyalty to the U.S."[24]

FBI agents interviewed von Braun at his office at NASA in Washington on November 4, 1970. Large sections of the interview report are blacked out and still classified "secret." While its contents are largely unknown, the interview report contains the following intriguing comment: "In 1947, he was allowed to return to Germany, and while there he married his present wife, who is also his cousin. They have been happily married for over twenty years, and since his marriage he has been a 'pretty good boy.' "

Had von Braun been accused of simultaneous infidelity to both his wife and country? Von Braun suggested that the source of the information that triggered the inquiry "may have told a 'cock and bull story' to American authorities in order to impress them."[25]

The FBI reported back to Butterfield at the White House the day after it interviewed von Braun. The report said that "he completely denied these allegations. He voluntarily furnished information concerning his activities since coming to the United States which would tend to discredit this allegation."[23]

As a resident of the United States for twenty-five years, a citizen for fifteen years, and serving United States government agencies for twenty-five years, Wernher von Braun was still not trusted or respected in some areas. But what could he do about it?

America's first—and to date only—space station began to take shape after the first successful landing on the moon. In form, it was quite modest when compared to Wernher von Braun's proposals of the 1950s, and in function, it had modest ambitions. On July 22, 1969, as *Apollo 11* returned to earth, NASA authorized its centers to begin work on the "Apollo Applications Program," which later took the name "Skylab."

For several years, NASA centers had been studying the possibility of orbiting a small laboratory in which scientific experiments might be conducted. Rival plans emerged from Marshall Space Flight Center in Huntsville and the Manned Spacecraft Center in Houston. Both plans used a Saturn V third stage (the S-IVB component) as the shell in which the orbiting habitat and workshop would be built. Marshall, while under the direction of von Braun, had favored a "wet" plan, in which a fueled—or wet—S-IVB would be launched into orbit on top of a Saturn IB first stage. A second Saturn IB would carry a crew into space to build the laboratory into the burnt-out orbiting S-IVB stage. NASA staff at Houston found this approach cumbersome and problematic in view of the amount of construction and assembly that would have to be done in orbit. Houston favored the "dry" approach, in which the entire satellite workshop would be assembled on the ground and launched into orbit as the third and final stage of a Saturn V rocket. In the end, Houston's "dry" approach won out.

As its design and mission emerged, Skylab would be the simplest of manned space stations. Like Apollo, it would have a crew of three, who, if all went according to plan, would not venture out of the satellite. Skylab would never serve as the staging base for a manned mission to Mars, but it might serve as a test site for equipment and men who would eventually take on that great adventure.[26,27]

The planning of the reusable launch vehicle had been under way for some time. The concept favored within NASA was a variant of von Braun's earlier designs for a winged vehicle that would be launched vertically and incorporate the knowledge gained from the Saturn V. The new reusable launch vehicle would be a two-stage vehicle with the orbiter riding piggyback on the booster. Both stages would have rocket and jet engines, and both would be manned. The booster would lift off under rocket power, then detach from the orbiter to fly back to base under jet power. The orbiter, or "space shuttle," as it eventually became known, would continue on into space under rocket power. The orbiter would be enormous—about the size of a Boeing 727 jet airliner—and it would carry its fuel internally. When its mission in orbit was accomplished, it would dip back into the atmosphere and fly back to base under jet power.[6,9,28] The outstanding element of this launch vehicle was that both booster and orbiter would be 100 percent reusable. This feature, it was proposed, would lower the cost of orbiting one pound of payload from $500 to $50, and would open new possibilities for the scientific and commercial exploitation of space.[6]

NASA's estimated cost for developing the completely reusable vehicle was from $10 to $13 billion,[29] operational expenses would be additional. Overall, the new launch vehicle would cost less than the Apollo program; but, by any way of reckoning, it would be an enormous expense. To bring the cost down to a more palatable amount, NASA director Thomas O. Paine proposed canceling the last four Apollo missions to the moon, resulting in an overall savings of $6 billion.[28] The launch vehicle was still too expensive for the Nixon administration and the Congress. NASA went back to the drawing board and Paine resigned in frustration.

The space transportation system that we know today was first presented to the American people on January 5, 1972, at a photo opportunity—not a press conference—by President Nixon and his new NASA administrator, James C. Fletcher, at the "Western White House" overlooking the Pacific Ocean in San Clemente, California. Nixon spoke briefly, announcing that he had decided that day to support development of the vehicle, although the real agreement between Nixon and Fletcher seems to have been made in late December 1971. During the photo opportunity, Fletcher held a model of the space shuttle as if it was zooming through the air like a conventional airplane. The President sat awkwardly, holding a model of the boosters and main fuel tank.[30] Nixon had agreed to budget $5.1 billion to develop the shuttle system.

The space transportation system, which Nixon approved and that we have today, is composed of an orbiter, the space shuttle; an external, expendable fuel tank; and two solid-fueled boosters, which are also reusable. Its major advantage—as seen by President Nixon in 1972—was that it would cost half as much as the 100 percent reusable system NASA originally proposed, though operating costs would be greater.

The faults of the space shuttle system were many, and Wernher von Braun viewed the final design as dangerous. Solid-fueled rockets had been used for manned spacecrafts only as part of emergency escape systems and braking rockets, never as boosters. If anything went wrong, they could not be shut off. Compounding the danger for the crew was the absence of an emergency escape system, which would have required an unacceptable increase in cost.[31] Prior to the space shuttle, astronaut safety had been a prime concern of NASA, and escape systems integral to the designs of all spacecraft. In fact, von Braun had designed an escape system for a very similar vehicle as part of the *Collier's* series on space exploration.[32] This omission would return to haunt NASA and the nation in the form of the tragedy of the space shuttle *Challenger*.

It would be fair to say that the design of the space shuttle and its booster system were dictated by Richard Nixon and the bean counters at the Office of Management and Budget, not by the rocket experts. Faced with the financial drain of the war in Vietnam, the declining interest in the space program, and a payoff that would happen long after he left office, President Nixon opted for the stripped-down model. Ironically, the final cost of developing the space shuttle system was $10 billion, about twice the amount Nixon approved.[33]

The tightening of NASA's budget by President Nixon with the support of Congress was soon felt in Huntsville in what the members of Wernher von Braun's rocket team called the "Great Massacre." According to civil service rules, nonveterans were the first to be released from their jobs. While many of von Braun's team could claim to be veterans of World War II, they had been soldiers in the wrong army. They were forced to take lesser positions within NASA or retire. They resented their treatment and conveniently ignored the fact that employees throughout NASA were given the same options.[34] Their semiofficial historians and authors of *The Rocket Team*, Frederick I. Ordway III and Mitchell R. Sharpe, had the appallingly bad taste to characterize the early retirements as "a final solution to the German problem."[35] Arthur Rudolph, who was the Saturn V program manager at Huntsville retired in late 1969 after the success of *Apollo 11*, and moved to San Jose, California.[36] Kurt Debus, director of the Kennedy Space Center and the man responsible for launching all Apollo flights to the moon, retired in November 1974.[37] Eberhard Rees, who had been von Braun's deputy director for decades and his successor as director of Marshall Space Flight Center, retired in January 1973.[38] By 1976, only eight of the original 118 members of von Braun's rocket team were still on staff at the Marshall Space Flight Center in Huntsville.[39]

For Wernher von Braun, his future was clear from the day NASA administrator Thomas O. Paine submitted his resignation. One of his German-born colleagues observed, "And from then on, von Braun became a non-person. . . . You could see him wandering up these long corridors all by himself, up and down. Then he left."[40]

Von Braun announced his resignation from NASA on June 10, 1972.[41] The space shuttle design was a disaster waiting to happen. The space station was deferred. The expedition to Mars was not likely to happen—not with Richard Nixon in the White House. Why stay around?

The timing of von Braun's resignation was better than he might have imagined. One week later, during the night of June 16–17, five men were arrested while attempting a burglary at the headquarters of the Democratic National Committee in the Watergate apartment and office complex in Washington, D.C.[42] Watergate entered the American vocabulary as a synonym for the criminal arrogance of the executive branch of the government. It would sap President Nixon's energy, preventing him from doing anything other than defending his doomed administration. It pushed news of the space program to the newspapers' back pages—if it even appeared there. With the first fifteen minutes of the evening television news devoted to the Watergate scandal, there was little time left for news of space. In any case, public interest had faded and the space program was winding down.

18

Exile in America: von Braun's Last Years

> To millions of Americans, Wernher von Braun's name was inextricably linked to our exploration of space and to the creative application of technology. . . . Not just the people of our nation, but all the people of the world have profited from his work. We will continue to profit from his example.
>
> President Jimmy Carter[1]

On June 30, 1972, Wernher von Braun cleared his desk at NASA headquarters, stepped into the revolving door that connected the federal government in Washington to American industry and emerged the following day, July 1, at Fairchild Industries in Germantown, Maryland. At the age of sixty, his blond hair turning to silver, he took on his first job in the private sector. His position at Fairchild's corporate headquarters was executive vice president for engineering and development.[2] Fairchild's president, Edward G. Uhl, had known von Braun since at least the late 1950s when he worked for the Glen L. Martin Company, the manufacturer of the Pershing missile, which von Braun's Group developed at Huntsville.[3] Uhl and von Braun were friends and hunting buddies. Uhl had moved on to become president of Fairchild, and when the position of corporate vice president opened up in 1971, he offered it to von Braun and held it open until von Braun was ready to move.

Fairchild Industries was a medium-size aerospace manufacturer. Since its creation in 1920, it had built a series of unexciting, yet stable aircraft that were used for photo reconnaissance, transporting military cargoes and troops, and commercial passenger transportation. In 1972, Fairchild had the A-10 Thunderbolt on its drawing boards. The A-10 was slow and stable enough in flight to mount antitank cannons. Fairchild was the primary contractor for several NASA satellites and supplied subsystems for Skylab. It also had a fair number of divergent operations that kept its focus pleasantly blurred, these included liquid waste treatment systems, land development, and three commercial radio stations.[4] Fairchild did not build rockets.

While von Braun's position with Fairchild came with a lot less influence and prestige than at Peenemuende and Huntsville, it was the best the world had to offer at the time. His new coworkers treated him with friendship and deference bordering on idolatry. Corporate headquarters was a short drive by company chauffeured limousine from his home in Alexandria, Virginia, and Washington, D.C., the political center of the Western world was still nearby. Fairchild's ongoing projects, while not revolutionary, were substantial, and von Braun was charged with strategic planning for the company's future.[5]

In the summer of 1973, von Braun allowed himself to have an extensive physical examination—it was Fairchild's company policy. An X ray taken as part of the exam showed a shadow near one of his kidneys. In September, he entered the Johns Hopkins Hospital in Baltimore, Maryland, where the surgeons removed his left kidney, which was found to have a malignant tumor. His surgery was followed by a course of radiation therapy, and when it was over, Wernher von Braun was as fit as ever. He had beaten cancer as he had triumphed over most challenges of his life.[6,7]

A major part of von Braun's job at Fairchild seems to have been the opening of doors for the company to powerful people in high places. He may not have been part of the power structure in Washington, but he was still a celebrity. And Washington has always liked celebrities. When Fairchild needed help, von Braun led the way to Capitol Hill, where he had spent years cultivating friends.[8] He arranged for Fairchild executives to meet the Shah of Iran and Prince Juan Carlos of Spain. On the latter occasion, the Spanish Air Force gave von Braun the mixed pleasure of sitting in the cockpit of a Messerschmitt 109 fighter,[2] just like the ones he had flown while serving in the Luftwaffe of Hitler's long defunct Third Reich.[9]

Von Braun's pet project at Fairchild appears to have been the company's

Applications Technology Satellite (ATS). The ATS was a high-powered television relay that facilitated reception on the ground with minimally expensive equipment; it required only a three-meter parabolic antenna and simple electronics. Von Braun viewed it as the ideal way to present educational and public service programming. The NASA-sponsored ATS was launched in 1974 and placed in geosynchronous orbit over India where it broadcast government-sponsored programming to 2,400 remote villages. Von Braun's and Fairchild's attempts to sell the same technology elsewhere did not catch on at the time. There was no money to be made by broadcasting educational programs to poor markets.[10,11]

In looking at Wernher von Braun's post-NASA-Fairchild years, one can ask, what new concepts did he generate? What programs did he initiate? What significant contributions did he make? One can find nothing truly innovative, nothing with his brand on it. The most significant activity is his support for direct television broadcasts, an idea which he did not create, yet which found commercial life decades later.

The great adventures in space were winding down and the fuel that had driven them, competition between the United States and the Soviet Union, was burning out. The last man to travel to the moon departed for earth on December 14, 1972.[12] The thirteenth and last Saturn V to fly lifted the Skylab space station into orbit on May 14, 1973.[13] Three crews of three astronauts each flew to Skylab aboard Saturn IB booster rockets. The last completed its mission on February 8, 1974.[14]

The manic thrust into space that had consumed the United States for fifteen years was stalling out, the victim of the Vietnam War that divided the country, growing détente with the Soviet Union, and ever-changing earthbound priorities. Realizing this, NASA administrator James C. Fletcher and his deputy, George Low, determined that they needed a grass-roots organization to demonstrate public support of the lagging space effort. They went to Edward Uhl of Fairchild Industries for help, and he assigned Thomas Turner, his vice president of marketing, to conduct a feasibility study. Three months later, Turner delivered a plan for a non-profit educational and scientific organization that was incorporated as the National Space Association on June 13, 1974. Fletcher and Low then brought pressure to bear among aerospace contractors and got about $500,000 to bankroll the new venture. (A year after its founding, its name was changed to the National Space Institute to avoid any inference that it was a lobbying group for NASA—which in fact it was. In 1986, it became the National Space Society.)[15,16]

To head the organization, Fletcher and Low needed a man of stature who had credibility in the field of space exploration, a master salesman who could resell the people of the United States on space. They went, of course, to Wernher von Braun. His first reaction was negative. "Another talking club," he called it. But they persisted, offering him a forum and a mission in life, which he seemed to have lost after he left Huntsville. He gave in and became the National Space Institute's first president.[15]

The membership of the group was far different from that of the VfR (*Verein fur Raumschiffahrt* or Society for Space Travel) that got von Braun and his youthful colleagues fired-up forty years earlier. The board of directors of the National Space Institute was composed of establishment politicians, aging entertainment personalities, and corporate executives. It filled its board of governors with cultural dinosaurs such as Senators Barry Goldwater and Hubert Humphrey, the presidents of the National Geographic Society and the Rockefeller University, television's Catholic evangelist Fulton J. Sheen, comedian Bob Hope, lesser celebrities from the entertainment industry, and a couple of ex-astronauts. As advisors it called upon a group of knowledgeable, retired NASA administrators.[16] The leadership of the National Space Institute likely inspired fear and loathing in American youth—which was still getting over the Vietnam War—as much as the desire to go to Mars.

The National Space Institute was created not to pursue some extravagantly impossible venture, but to build an organization, a stable group of people who could safely share their interest through meetings in hotel conference rooms and glossy publications. It was a forum in which Wernher von Braun and his disciples could pursue his aging dream without taking chances. The effectiveness of the group in inspiring space exploration is attested to by the fact that since its founding in 1974, no human being has gone into space beyond earth orbit, and there are no plans by any government or group to do so.

Von Braun spent forty years of his life pursuing goals that needed the help and the support of others. By the mid-1970s, he needed the support of the American people and a large block in Congress to push his plans for space exploration farther than the moon, and it looked like he was not going to get it. Possibly as a way of dealing with the frustration, of achieving something by himself without the need for others, and also for the sheer joy of it, he returned to an old love. After a forty-year intermission, he again took up flying sailplanes. While in Huntsville, von Braun had obtained a pilot's license for flying small, single-engine aircraft.[17] Subsequently, he earned a commercial license,[18] which allowed him to slide behind the controls of larger aircraft carrying delegations to and from Huntsville.[19] After joining Fairchild, he earned certification for flying seaplanes.[6]

But sailplanes are different. Sailplanes, or gliders, are airplanes without engines. They are highly efficient machines for catching air currents and, like buzzards and birds of prey, riding them upward and over distances.

All that keeps them aloft are invisible air currents and their pilots' ability to use them.

As a teenager, Wernher von Braun did his first flying in gliders above the Galgenberg mountain in his native Silesia. Now, on Saturday mornings, he drove to Cumberland, Maryland, where he once again matched his skills with the rising air currents as he glided above the Appalachian Mountains.[8] The sport of soaring has its levels of accomplishment, and von Braun demonstrated his proficiency and earned a Silver Badge by soaring to an altitude of 11,000 feet above the Adirondack Mountains of upstate New York.[6] The Silver Badge was von Braun's alone; he did not have to ask anybody for their support or sell anybody anything to get it.

On July 15, 1975, the last functional Saturn IB was scheduled to carry three American astronauts into orbit where they would link-up with two cosmonauts in a Soviet Soyuz spacecraft. Competition had given way to tentative cooperation in space.[14] Wernher von Braun was at the Kennedy Space Center to observe the launch of the Apollo half of the Apollo-Soyuz test program. Von Braun was thrilled by the successful launch and no doubt proud that his Saturn boosters had a perfect record, not a single Saturn I, Saturn IB, or Saturn V ever failed to complete its planned flight. Not one had ever blown up on the pad or in flight, a record unmatched by any other rocket design.[20]

When the three astronauts were safely in orbit, von Braun quietly boarded an airplane for Stuttgart, Germany. There he accepted a seat on the board of directors of Daimler Benz, the manufacturer of the Mercedes Benz.[21] There was still a market for his skills as an engineer and his talents as a visionary.

Soon thereafter, in the summer of 1975, von Braun was on vacation with his family in Ontario, Canada, when one morning he discovered that he had passed a trickle of blood, but the sign of intestinal bleeding soon passed, and he was willing to deny to himself that it had even happened. A few weeks later, while in Alaska on a business trip with his boss, Ed Uhl, the symptom returned. Von Braun checked into the Johns Hopkins Hospital on August 6. A thorough physical exam indicated a malignant tumor of the large intestine. The surgeons resected the colon, removing the tumor, and expected their patient to be discharged within four weeks. Von Braun lived exclusively on intravenous fluids for several weeks and developed a high fever. He was discharged from the hospital on September 29, seven weeks after he entered, twenty pounds lighter and noticeably weakened.[6,22]

Wernher von Braun returned to his office in November 1975, but under doctors' orders, he put in short hours and no longer placed long range commitments on his calendar. His physical condition continued to deteriorate. He had an infected intestinal abscess that caused a persistent fever and his colon began to bleed. He was back in the hospital again in May 1976 to receive transfusions and intravenous feeding. This stay was at a hospital in Alexandria, Virginia, farther from the specialists, but closer to home. There would be no more surgery and no cure for his cancer.[22]

Von Braun was in and out of the hospital in Alexandria after that, as his body wasted away. By his own estimate, he received over 400 units of blood, but, he pointed out proudly, it was "American blood."[23] He accepted the inevitable and retired from Fairchild Industries on December 31, 1976.[24]

In early 1977, President Gerald R. Ford awarded Wernher von Braun the National Medal of Science. His condition prevented him from making the short trip across the Potomac to receive it personally. His health had deteriorated so much that his doctors refused to even let White House staff members visit him. Von Braun's close friend and former boss Ed Uhl presented the medal to him in the hospital. Von Braun was truly touched to receive the high honor and sign of appreciation from his adopted country.[24,25]

In Wernher von Braun's final days, his family gathered in Alexandria to help him through the pain. Maria, his wife of thirty years, and Peter, his seventeen-year-old son were there. His daughter Iris, now twenty-eight, who had married an Indian businessman, returned from New Delhi. His twenty-five-year-old daughter Margrit, who was married and living with her husband in Idaho, also returned.[26] The odyssey of the rocketeer ended quietly in an Alexandria, Virginia, hospital on June 16, 1977, sixty-five years after it began.

19

Epilogue: The Man Who Sold the Moon

> Wernher von Braun plays an absolutely central role in the history of rocketry and in the development of space flight—equally on the inspirational as on the technological side.
>
> On the other hand, . . . I found many things deeply disturbing about his career—his willingness to work for the Wehrmacht and the SS, to accept a commission in the SS, to brief Hitler and Himmler, and to use slave labor in the production of the V-2s.
>
> Carl Sagan[1]

Wernher von Braun was buried thirty-two years after the defeat of Nazi Germany. Survivors of the war had gone on to new lives. Memories faded. A whole generation had been born without memories of the Third Reich. The Nazis and their atrocities were forgotten, except by their victims.

The Mittelwerk and the Dora concentration camp were virtually unknown to the historians of rocketry and the space age and to students of the crimes of Nazi Germany. It should not be too surprising that the Mittelwerk and Dora are absent from most histories of rocketry. These accounts were written by Wernher von Braun, Walter Dornberger, and members of von Braun's German rocket team. One of the big questions never answered by these authors is how and where Germany built six thousand V-2s, half of which were fired in combat. One might have expected Jewish activists like Simon Wiesenthal, to have kept alive the memory of Dora. But they were as much in the dark as everyone else. Few Jews ever got to Dora; they were sent to the extermination camps in the East. Dora's prisoners were more often than not political prisoners who had technical skills that were useful in building rockets. Knowledge of the Dora concentration camp, the Mittelwerk, and the crimes committed there were suppressed by factions in the United States Army; and other parts of the federal government had little interest in digging through the remnants of history.

Despite the efforts of the United States government and the Germans brought to this country by Project Paperclip to keep the Mittelwerk a secret, the ugly facts emerged on a few occasions. In 1964, American historian and former Allied intelligence officer James McGovern published his book *Crossbow and Overcast*, which recounted some of the V-2's history and the hiring of von Braun and the German team. McGovern told of the liberation of Dora and the Mittelwerk, the sickening conditions U.S. forces found there, and the hundreds of emaciated corpses waiting their turn in the crematoria.[2] The Friends of Deportees of the Dora-Ellrich camps protested in print in 1965 and 1966. But that was in France, where it was invisible to most Americans, and Wernher von Braun denied any involvement or responsibility.[3] Charles Lindbergh, the aviation pioneer, visited Dora and the Mittelwerk soon after its liberation, and he wrote with great moral outrage about the inhumane conditions he saw there in *The Wartime Journals of Charles A. Lindbergh* (1970).[4] The puzzle here is that Lindbergh knew von Braun and met him at the Kennedy Space Center at the time of the *Apollo 11* launch in July 1969. Lindbergh must have been preparing his book for publication at the time, but there is no evidence that he asked von Braun about the factory that built his V-2 rocket. It was the pattern everyone followed, the path everyone took. Von Braun so successfully distanced himself from his Nazi partners that nobody made the connection or asked the question: What did Wernher von Braun know about war crimes at Dora and Mittelwerk, and was he involved?

The Nazis are remembered best by the Jews. It became a cultural axiom that they will never forget the Holocaust. Yet it was not until the late 1970s that Jewish American political-action groups convinced Congress that there were Nazi criminals who had escaped punishment living in the United States, and that these criminals should be found and exposed.

The crusade against Nazis living in the United States was led by Congresswoman Elizabeth Holtzman of New York. In the latter half of the 1970s, she chaired a House Judiciary Subcommittee set up to investigate the government's role in bringing Nazis into the United States and to expose those who had entered the United States fraudulently. Her effort culminated in 1978 with passage of the Holtzman Amendment to the United States immigration law, which provided the legal basis for deporting Nazi war criminals and for refusing entry of Nazi war criminals. In addition, Holtzman's investigation also embarrassed the executive branch into taking action of its own. In 1979, the Justice Department established the Office of Special Investigations (OSI) to track down Nazi war criminals.[5] The OSI went to work with a staff of four investigators, seven historians, and about two-dozen lawyers.[6]

While America took leisurely and late steps toward justice, a Frenchman named Jean Michel, who had been a member of the Resistance and who had been imprisoned in the Dora concentration camp, wrote *Dora*, his memoirs about the camp. The book was first published in France in 1975; the English translation was not released in the United States until 1979. Additional text on the book's cover expanded on the apparently benign title: "The Nazi concentration camp where modern space technology was born and 30,000 prisoners died."[7]

Michel wrote passionately about his two years in Dora, from his arrest by the Gestapo in Paris for being a member of the French Resistance to his liberation when he was near death, after he had been used and wasted as a slave by the Third Reich.

It is hard to phrase moral outrage more powerfully than did Jean Michel in his introduction. He wrote:

> Auschwitz, Treblinka, Buchenwald, Dachau: any of the men responsible for these infernos who escaped punishment will be damned for eternity. Dora is different.
>
> Men who were intimately involved with the creation and operation of the camp are today respected, venerated and admired. . . . [T]hey have drawn a veil over the indisputable and atrocious fact that this slavery, this unspeakable sum of suffering, misery and death, was put, at Dora, to the service of the manufacture of missiles . . . which later—after the Russians and the Americans had shamelessly scooped up the scientists of the Reich . . . —made possible the conquest of space.[8]

Michel vividly remembered two "official visits of inspection," which were preceded by increased discipline of prisoners including a doubling of the beatings they received. The first inspection, on December 10, 1943, was made by Hitler's minister of armaments, Albert Speer.[9] Speer later claimed that he was severely shaken by seeing the unbearable living conditions and the deplorable physical condition of the prisoners.[10] The second inspection took place on January 25, 1944.[9,11] While Michel vividly remembered the effect of the visit on the prisoners, he did not learn until 1967 that the official visitor was Wernher von Braun. Michel generously gave von Braun the benefit of the doubt. He did not hold von Braun responsible for the crimes committed at Dora and the Mittelwerk, but for creating a monstrous distortion of history by not telling what he saw there, and of being a conscienceless mercenary because he worked for the Nazis.[12]

Early in 1980, Harvard law student, Eli Rosenbaum, who later worked for the OSI, picked up a copy of *Dora* in a Cambridge bookstore. "I was stunned," Rosenbaum said later. "I had never seen camp Dora mentioned in any Holocaust literature." The Jewish community had been thorough in investigating concentration camps where their people had been imprisoned and died, but, understandably, since Dora was not created as part of Hitler's Final Solution, they had passed over it. Rosenbaum's search for more information about Dora led him to read *The Rocket Team* by Frederick I. Ordway III and Mitchell R. Sharpe. The authors, close friends of Wernher von Braun and his team of German rocket scientists, disingenuously reported that V-2 production at the Mittelwerk was supported by prison labor, but they made no mention of Dora concentration camp and slave laborers. Rosenbaum was appalled to read that the production manager, Arthur Rudolph, had complained bitterly about having to leave a New Year's Eve party (January 1, 1944) to solve a production problem—Rudolph expressed no concern for his slave laborers who worked under inhumane conditions while he partied. When Rosenbaum went to work for OSI, he asked the deputy director if he knew about Rudolph. He did not, but he authorized Rosenbaum to start collecting information.[13]

The most damning evidence against Arthur Rudolph was testimony and depositions given at the Dora-Nordhausen War Crimes Trials in 1947,[14] which finally surfaced at the National Archives in 1981. Witnesses testified that Rudolph had the responsibility of requisitioning slave laborers from the SS, allocating the criminally inadequate food supply for prisoners, and submitting accusations of sabotage by named prisoners to the SS for punishment. There were also accusations that Rudolph and his subordinates were involved in the punishment and execution of prisoners.[14,15]

Staff members of the OSI tracked down Arthur Rudolph to his retirement home in San Jose, California, and questioned him about Dora and the Mittelwerk on October 13, 1982, and again on February 4, 1983.[15] They returned again late in 1983 to offer Rudolph a deal: Leave the country and renounce his citizenship, or face trial and lose his NASA pension.[16] Rudolph chose exile to his native Germany. His reasoning: "I didn't fight it because it would have dragged on for years. I am an old man."[17] He was seventy-six at the time.

Arthur Rudolph signed his agreement with the OSI on November 28, 1983. He boarded a Lufthansa flight for Germany on March 27, 1984, and surrendered his United States citizenship on May 25, 1984. The OSI finally got around to announcing the details of the Rudolph case on October 17, 1984, eleven months after Rudolph agreed to return to Germany.[18] If it was such a big victory for justice, why did the OSI take so long to claim it? Why didn't the OSI bring Rudolph to trial on the basis of the evidence against him? Why didn't the OSI reveal the complicity of government agencies, especially the Army, in covering up the Nazi pasts of Rudolph and others who immigrated under Project Paperclip? "Teaching moral lessons is not what Congress set us up for," said a representative of the OSI. "It's the ultimate aim of this office to deport people who persecute people."[6] The OSI continued its search for other geriatric Nazis.

Arthur Rudolph's fate was felt most strongly in Huntsville, Alabama. More than fifty of the original 118 German rocket experts still lived there. They were now known as the "Old German Team." Their first reaction was to fight their colleague's treatment. It was not fair. The United States had warmly accepted Arthur Rudolph nearly forty years earlier, and had used his talents over a long period of time. Now, after the success of *Apollo 11*, after his retirement in 1969, a bunch of zealots dredged up the long-dead past and threw him out of the country—his country. The zealots could do the same to them, to the last of the Old German Team. The Justice Department estimated that between ten and fifteen of them had been there, working beside Rudolph at the Mittelwerk. The comments of those who talked to the press were equivocal:

"You had to be a [Nazi] party member to get up in life, like the Kiwanis here."

"We all worked there [the Mittelwerk]. We didn't fight it."

"People are concerned because they fear a lot of dirty stuff could be dredged up and put on the table. Any prisoner could pop up and say, 'You there! You beat me.' "

"What happened to Arthur could happen to any of us."

For the most part, however, they hid behind closed doors, frightened old men who wanted to live out their last days in peace and forget the crimes of the past.[17]

NASA officials, especially at Marshall Space Flight Center in Huntsville, did not want to get involved. Arthur Rudolph was history, and nobody remembered him well enough to comment.

At the Alabama Space and Rocket Center (now the United States Space and Rocket Center), which von Braun helped to found, its director, Ed Buckbee, who had been public affairs spokesman for the Marshall Space Flight Center, said, "The Army investigated all these guys when they came in. Why are they pursuing this now?"[17]

John Mendelsohn, historian of the Nazi era for the National Archives, tried to understand the contradictions. "What do you do with these people?" he asked. "Here you have these people, fallen angels maybe, who are directly responsible for us getting to the moon before the Russians. At the same time, these are people with clay feet who are responsible for untold misery. Yet when you question them, they don't even see it."[17]

The OSI investigated other living members of the German rocket team, but took no action against any of them.[19] It did not investigate von Braun with the same vigor it pursued Rudolph or the others; von Braun had passed beyond the OSI's jurisdiction a few years earlier. Research on von Braun's activities during World War II was left to independent journalists and historians.

Large sections of the dossier on Wernher von Braun kept by United States Army Intelligence were declassified on November 27, 1984,[20] a month after the Rudolph case was made public knowledge. His FBI dossier was declassified on February 28, 1985.[21] The documents in these files, along with the transcript of the Dora-Nordhausen War Crimes Trials,[14] gave a whole new perspective on the German American hero. They told of his membership in the Nazi party, his rank of major in the SS, his association with the Dora concentration camp, and the cover-up of these facts to facilitate his entry into the United States.

Other bits of information have surfaced from other sources that implicate von Braun in war crimes. The OSI uncovered a transcript of a meeting held on May 6, 1944, at the Mittelwerk to discuss ways of increasing V-2 production. Participants in the meeting discussed bringing in 800 skilled French civilians. Less than a month after the meeting, over a thousand Frenchmen arrived at the Mittelwerk; more than 700 died there. The May 6 meeting on increased V-2 production was attended by high-ranking SS officers, Wernher von Braun, and Arthur Rudolph. Three others at the meeting who came to the United States with von Braun as part of the rocket team were Ernst Steinhoff, Hans Lindenberg, and Hans Friedrich.[22,23]

An even more damning piece of evidence implicating Wernher von Braun in war crimes is a letter uncovered by Michael Neufeld, Curator of World War II history at the National Air and Space Museum. The letter had been with documents in the possession of United States agencies for half a century. The letter was written by Wernher von Braun to Albin Sawatzki, the production planner of the Mittelwerk, about his efforts to find personnel to staff a workshop to evaluate "ground vehicle test devices." Von Braun wrote:

> During my last visit to the Mittelwerk, you proposed to me that we use the good technical education of detainees available to you at Buchenwald to tackle . . . additional development jobs. You mentioned in particular a detainee working in your mixing device quality control, who was a French physics professor and who is especially qualified for the technical direction of such a workshop.
> I immediately looked into your proposal by going to Buchenwald . . . to seek out more qualified detainees. I have arranged their transfer to the Mittelwerk with Standartenfuehrer [Colonel] Pister [Buchenwald Camp Commandant], as per your proposal.[24]

Not only does von Braun's letter show him conducting business at two concentration camps, it also demonstrates in his own words that he recruited slave laborers to build his V-2 rockets. According to the principles followed at Nuremberg, doing so was a war crime.

Now that it is all over, Arthur Rudolph seems to have been the scapegoat for war crimes committed by Wernher von Braun and his German rocket team. Rudolph, because of his position at the Mittelwerk, along with bad luck, personal longevity, and testimony against him, was the only one of the group seriously investigated for war crimes. Regrettably, the OSI's investigation of the team was cursory, the case plea-bargained, the results published belatedly, and the punishment clearly not fit for the alleged crimes. Though he tried to regain his citizenship and be absolved of the accusations against him, Arthur Rudolph died in Germany on December 29, 1995, at the age of eighty-nine, an exile in the land of his birth.[25]

Contention over the memory of Wernher von Braun and his fellow rocket builders reached back in time and geography to Germany. The German Aviation, Space and Armaments Industries Association of a reunited Germany planned a celebration for October 3, 1992, at Peenemuende on Germany's Baltic coast to celebrate the fiftieth anniversary of the first successful firing in 1942 of the A-4/V-2 rocket, an event viewed by many as the beginning of the space age. The event evoked a public outcry that made front-page news around the world. Those who protested the celebration remembered the devastation caused by the rocket in the bombardment of England and the Low Countries nearly a half century earlier, and they knew that those who actually manufactured the V-2 were slaves in Nazi concentration camps. German defense minister Volker Ruehe characterized the planned celebration as "tactless," and he added, "You cannot isolate technology from history." The celebration was ultimately canceled at the request of an embarrassed Chancellor Helmut Kohl.[26]

The end of World War II, the destruction of the Third Reich, the suppression of evidence by a faction within the United States Army, and the attrition of witnesses over time conspired to conceal the facts of Wernher von Braun's life in Nazi Germany. Yet, the myth that he had so carefully woven about his life inevitably unraveled. There were too many witnesses and too many documents that bore his name for all of the evidence to remain buried forever. While many details have been lost, the most damning facts are now known: Wernher von Braun willingly joined the Nazi cause and was an accomplice in its crimes in order to build rockets and to promote his own career.

When the Third Reich collapsed, von Braun unashamedly switched his allegiance to the victor. He became a leader in the United States' space program and the prophet of space travel. Because of his tireless promotion, he was the man who sold the moon. Sadly, because of his complicity with the Nazi cause, he also sold his soul to reach that goal.

Abbreviations to References

Abbreviations for the most often cited archival and related sources are:

Army FOI: United States Army Intelligence and Security files on Wernher von Braun obtained through a Freedom of Information Act request, 727 numbered pages.

FBI FOI: FBI files on Wernher von Braun obtained through a Freedom of Information Act request, 532 pages, not numbered or indexed.

USA v. Andrae: *United States of America v. Kurt Andrae et al.*, National Archives Microfilm M1079.

von Braun Bibliography: "A Bibliography of Wernher von Braun with Selected Biographical Supplement, 1930–1969," compiled by Mitchell R. Sharpe, NASA; www.msfc.nasa.gov/general/history.html

von Braun Papers: The papers of Wernher von Braun, Library of Congress Manuscript Division. These papers are organized into 61 containers.

von Braun Resume: Resume of Wernher von Braun, dated January 1967, obtained from NASA.

Notes

Complete reference information for works identified by author's name or abbreviated titles can be found in the Bibliography. Internet sources are identified by their standard prefix, www. Information found in encyclopedias and standard histories is cited as being from "Common reference sources."

PRELUDE: THE ROCKET

1. Churchill, 38.
2. Irving, 285–86.
3. Ley, *Rockets*, 216–17.
4. Dornberger, 6–8. Details of V-2 launch.
5. Irving, 290.
6. Ordway and Sharpe, 195.
7. Irving, 295.
8. Churchill, 52–53.

CHAPTER 1: THE IMMIGRANT

1. Army FOI, 502. Qualification Sheet for German Scientific Personnel, von Braun, Prof. Dr. Wernher, dated June 10, 1945.
2. Ibid., 262. Information sheet on von Braun, no heading or date.

3. Ordway and Sharpe, 310–12.

4. von Braun and Ordway, *History*, 122.

5. Army FOI, 513–517. Basic Personnel Record of von Braun dated September 20, 1945.

6. McGovern, 94–95.

7. "Marshall's Roots," www.msfc.nasa.gov/general/history.html

8. Bergaust, *Wernher von Braun*, 124–25.

9. Ibid., 125–26.

10. von Braun, Wernher, "Why I Chose America," *The American Magazine*, July 1952, 15, 111–12, 114–15.

11. von Braun Papers, container 47. von Braun, Wernher, manuscript titled, "Where Are We Going?"

12. Common reference sources.

13. Dornberger, 155–68.

14. Huzel, 133–42.

15. Ibid., 220.

16. Ibid., 214, 216.

17. Ordway and Sharpe, 317.

18. Dornberger, 26–27.

19. Bergaust, *Wernher von Braun*, xiii–xiv. Pages list all 118 men.

20. Dornberger, 30.

21. Army FOI, 415. Memo, Subject: Scientific and Technical Cross-reference: Personnel, dated April 30, 1947.

22. Ordway and Sharpe, 188.

23. Huzel, 117.

24. McGovern, 245.

25. Stuhlinger and Ordway, *Biographical Memoir*, 72.

26. Ibid., 26.

27. Ley, *Rockets*, 506.

28. Ordway and Sharpe, 354.

29. Ibid., 355–56.

30. Army FOI, 438. Distribution list "B."

31. Ordway and Sharpe, 71, 271–72.

32. Huzel, 215.

33. Ibid., 223.

34. Ibid., 220–21.

35. "German Scientists Tell," *El Paso Herald-Post*, December 5, 1946; copy in Army FOI, 245–46.

36. "V-2 Tests 33% Successful," *El Paso Times*, December 6, 1946; copy in Army FOI, 243.

37. V-2 Rocket sets New Speed Record," *El Paso Times*, December 6, 1946; Army FOI, 239.

38. "German Scientists Have Own Court," *El Paso Herald-Post*, December 6, 1946; Army FOI, 244.

39. Stuhlinger and Ordway, *Biographical Memoir*, 84–85.

40. FBI FOI, El Paso file no. 77–594 dated September 25, 1948; page 7 of 12.

41. "We Want with the West," *Time*, December 9, 1946, 67–68, 70.

42. von Braun Papers, container 46. "General Summary of Professor von Braun's Speech at the Rotary Club," January 16, 1947.

43. Ibid., container 58. Letter from Gordon L. Harris to Colonel E. W. Richardson, dated May 16, 1959.

44. Army FOI, 244. List of articles in El Paso *Herald-Post*.

45. Army FOI, 141–42. Request for Inclusion of Fiancee in Shipment of Families, signed H. N. Toftoy, November 26, 1946, and Request to Include My Fiancee in the Shipment of Families, signed Dr. von Braun, November 7, 1946.

46. Bergaust, *Wernher von Braun*, 551.

47. Ibid., 89–90, 94.

48. Stuhlinger and Ordway, *Biographical Memoir*, 84.

49. Lomax, 181.

50. Bergaust, *Wernher von Braun*, 105.

51. Reitsch, Hanna, 220–37.

52. Office of the United States Chief of Counsel for Prosecution of Axis Criminality, 551–71.

53. Army FOI, 190–91. Formal application in German by von Braun for permission to marry to *Reichsfuehrer SS, Rasse und Siedlungshauptamt*, dated April 5, 1943. As an SS officer, von Braun was required to get RuSHA's authorization prior to marriage; see also Krausnick et al., 597.

54. Army FOI, 416–17. Letter to Director of Intelligence from H. N. Toftoy, dated February 11, 1947.

55. Ibid., 321. Letter from Robert A. Schow to Headquarters, European Command, Intelligence Division, dated September 28, 1948.

56. Ibid., 727. Agent Report re. von Braun, dated December 3, 1954.

57. David, 112.

58. Army FOI, 371. Letter from Bruce W. Bidwell to Commanding General, Fourth Army, dated July 9, 1948.

59. Stuhlinger and Ordway, *Biographical Memoir*, 84.

60. Army FOI, 408. Letter from Chief, Security Group, to Director FBI, dated June 30, 1947.

61. "German Pupils Sing 'Eyes of Texas,' " *El Paso Herald-Post*, August 5, 1947; copy in Army FOI, 247–48.

62. " 'Speak English' Contests," *El Paso Herald-Post*, August 6, 1947; copy in Army FOI, 249.

63. "Scientists Not Seeking Citizenship," *El Paso Times*, July 27, 1947; copy in Army FOI, 240.

64. "German V-2 Scientists Here," *El Paso Times*, November 5, 1947; copy in Army FOI, 238.

65. McGovern, 247.

66. McGovern, 248.

67. Huzel, 221.

68. "German Scientists in El Paso Blasted," *El Paso Times*, July 1, 1947, 1; copy in Army FOI, 241.

CHAPTER 2: THE AUTHORIZED BIOGRAPHY

1. von Braun, Wernher, "Why I Chose America," *The American Magazine*, July 1952, 15, 111–12, 114–15.

2. von Braun, Wernher, "Reminiscences of German Rocketry," *Journal of the British Interplanetary Society*, May 1956; reprinted in Clarke, 33–55.

3. von Braun, Wernher, "Space Man—The Story of My Life," *The American Weekly*, July 20, 1958, 7–9, 22–25.

4. von Braun and Ordway, *History*.

5. von Braun and Ordway, *Rockets' Red Glare*.

6. Bergaust, *Wernher von Braun*, 549–51.

7. von Braun and Ordway, *History*, 64–65.

8. Stuhlinger and Ordway, *Biographical Memoir*, 16.

9. Ordway and Sharpe, 93–94.

10. Oberth, Hermann, "Hermann Oberth: From My Life," *Astronautics*, June 1959. Reprinted in Clarke, *The Coming of the Space Age*, 113–21.

11. Ley, *Rockets*, 114–20.

12. Ibid., 123–24.

13. Ibid., 126–27.

14. Ibid., 128–40.

15. Ibid., 143–44.

16. von Braun and Ordway, *Rockets' Red Glare*, 138.

17. Garlinski, 4.

18. Dornberger, 26–28.

19. Bar-Zorah, 17.

20. "Dornberger, Walter R(obert)," *Current Biography*, 1965, 125–27.

21. Shirer, 183.

22. Ibid., 195–200.

23. Dornberger, 31–36.

24. von Braun, "Konstructiv, theoretische und experimentalle Beitraege zu dem Problem der Fluessigkeitsrakete," (Ph.D. diss., Der Friedrich-Wilhelms-Universitat zu Berlin, 1934); copy in von Braun Papers, container 52.

25. Bergaust, *Reaching for the Stars*, 55.

26. von Braun and Ordway, *History*, 49–50.

27. Dornberger, 45–46.

28. Ibid., 47–49.

29. Ley, *Rockets*, 181–182, 193 (map).

30. Dornberger, 53.

31. von Braun Resume.

32. Dornberger, 42–46, 54–56.

33. Ordway and Sharpe, 30.

34. Dornberger, 64–66.

35. Shirer, 597–99.

36. Dornberger, 58, 60–63.

37. Ibid., 69.

38. Ethell, 41.

39. Ley, *Rockets*, 200.

40. Irving, 21.

41. Ley, *Rockets*, 200, 500.

42. Dornberger, 3–17.

43. Ordway and Sharpe, 42.

44. von Braun and Ordway, *Rockets' Red Glare*, 147.

45. Shirer, 925–32.

46. Dornberger, 98.

47. Ibid., 99–102.
48. Speer, *Inside the Third Reich*, 368.
49. Dornberger, 102–6.
50. Ibid., 107.
51. McGovern, 13, photo and caption after 120.
52. Irving, 103–15.
53. McGovern, 14–15.
54. Dornberger, 155–68.
55. Ibid., 185.
56. Ibid., 200–7.
57. Speer, *Inside the Third Reich*, 371–72.
58. Persico, 103.
59. "Stauffenberg, Claus Schenk Graf von," Wistrich, 298–300.
60. Shirer, 1044–45, 1048–69.
61. Speer, *Infiltration*, 213.
62. Ordway and Sharpe, 53–54.
63. Ibid., 193–94.
64. Hoess, 233–39.
65. Hilberg, 327.
66. Speer, *Infiltration*, 219.
67. Huzel, 119.
68. Dornberger, 141–43, 250–51.
69. Ordway and Sharpe, 254–55.
70. Huzel, 133–34.
71. Huzel, 145.
72. Dornberger, 266.
73. McGovern, 94–96.
74. Ordway and Sharpe, 184.
75. Ibid., 193–94.
76. Huzel, 149.
77. Ordway and Sharpe, 264–65.
78. Dornberger, 271.
79. Speer, *Infiltration*, 245.
80. McGovern, 130–32.
81. Ibid., 132–35.
82. Huzel, 187–89.
83. McGovern, 147.
84. Army FOI, 135. Memo from Headquarters, 66th CIC Group, U.S. Army, Europe, to Assistant Chief of Staff, G-2, Intelligence, U.S. Army, Europe, dated July 1953.
85. McGovern, 151–76.
86. Ibid., 197.

CHAPTER 3: THE COVER-UP

1. Ordway and Sharpe, 10–11, quote from U.S. Army officer who held von Braun and his team in custody immediately after they surrendered.

2. Huzel, 225.

3. Ordway and Sharpe, 292.

4. Hunt, Linda, "U.S. Coverup of Nazi Scientists," *Bulletin of the Atomic Scientists* (April, 1985): 16–24.

5. Army FOI, 168. Memo from Kurt Rosen, OMGUS for Germany to Capt. Hirsch. This document is reprinted in the Illustrations.

6. Piszkiewicz, 94–97. Himmler met von Braun at Peenemuende on June 28, 1943, the day von Braun was promoted to Sturmbannfuehrer.

7. Neufeld, photo after 210. Taken in June 1943, the photo caption identifies von Braun wearing an SS uniform in the company of Himmler and several German Army officers.

8. Army FOI, 170–72. Affidavit of Membership in NSDAP of Prof. Dr. Werner von Braun, signed and dated June 1, 1949.

9. Ibid., 492–95, OMGUS Revised Security Report re. Wernher von Braun dated September 18, 1947. Reprinted in Hunt, "U.S. Coverup of Nazi Scientists."

10. Ibid., 484–87, OMGUS Revised Security Report re. Wernher von Braun dated February 26, 1948. Reprinted in Hunt, "U.S. Coverup of Nazi Scientists."

11. *National Archives Microfilm Publications Pamphlet Describing M1079* (USA v. Andrae). (Washington: National Archives, 1981).

12. Ordway and Sharpe, 66.

13. Michel, 41–43, 62–63, 67.

14. Speer, *Infiltration*, 210–11.

15. Agoston, 27–29.

16. McGovern, 119–22.

17. Hunt, Linda, "Arthur Rudolph of Dora and NASA," *Moment* (April 1987): 32–36.

18. Hunt, *Secret Agenda*, 70–71.

19. USA v. Andrae. Defense Exhibit 38A.

20. USA v. Andrae. Prosecution Exhibit P-125A.

21. Piszkiewicz, 225–26.

22. USA v. Andrae.

23. Army FOI.

24. Ordway and Sharpe, 10–11.

25. von Braun and Ordway, *History*, 107, 116–18, 140.

CHAPTER 4: THE REBORN ROCKETEER

1. von Braun, *The Mars Project*, viii.

2. Bergaust, *Wernher von Braun*, 130.

3. McGovern, 210.

4. Ibid., 154.

5. Ibid., 209.

6. Stuhlinger and Ordway, *Biographical Memoir*, 74.

7. Ordway and Sharpe, 351–53.

8. Stuhlinger and Ordway, *Biographical Memoir*, 83.

9. *Postage Stamps of the United States*, (Washington, D.C.: U.S. Government Printing Office, 1953): 158.

10. Army FOI, 250–51. Letter of Identification for W. von Braun dated January 1, 1948; 721–25, memo for record dated July 29, 1958.

11. von Braun Resume.

12. von Braun, Wernher. "Why I Chose America," *The American Magazine*, July 1952, 15, 111–12, 114–15.

13. "von Braun, Wernher," *Current Biography*, 1952, 607–9.

14. von Braun, *The Mars Project*, 3.

15. Bergaust, *Wernher von Braun*, 155–56.

16. Ley, Willy. *The Conquest of Space*. (New York: Viking Press, 1949).

17. Clarke, "Epilogue," in Blueprint for Space, ed. Frederick I. Ordway III and Randy Liebermann, 218–19.

18. von Braun and Ordway, *History*, 125–26.

19. Bergaust, *Wernher von Braun*, 147.

20. Stuhlinger and Ordway, *Biographical Memoir*, 158.

21. Army FOI, 311. "Memo for Record," by W. H. B. dated June 30, 1949.

22. Ordway and Sharpe, 258–59.

23. Army FOI, 94. Memo from CIC Special Agent Robert J. Cannon to Officer in Charge, CIC Headquarters, dated April 19, 1948.

24. Army FOI, 95. Memo from James P. Hamill to Chief of Ordnance, dated July 22, 1948.

25. FBI FOI. Many documents relating to the initial investigation of von Braun.

26. Ordway and Sharpe, 359.

27. Stuhlinger and Ordway, *Biographical Memoir*, 85.

28. Army FOI, 123–26. Security clearance request by von Braun dated Oct. 15, 1954; 587, CIC Agent Report summarizing von Braun's naturalization records, dated April 20, 1955.

29. Ordway and Sharpe, 361.

CHAPTER 5: REDSTONE

1. von Braun, Wernher. "Why I Chose America," *The American Magazine*, July 1952, 15, 111–12, 114–15.

2. Common reference sources.

3. Bergaust, *Wernher von Braun*, 184–88.

4. "An Organizational History of the Redstone Arsenal Complex." www.redstone.army.mil/history

5. Bergaust, *Wernher von Braun*, 152–53.

6. von Braun and Ordway, *History*, 127.

7. von Braun Resume.

8. Ordway and Sharpe, 365–69.

9. Stuhlinger and Ordway, *Biographical Memoir*, 118.

10. Common reference sources.

11. Ordway and Sharpe, 370.

12. Gunston, 37.

13. von Braun and Ordway, *History*, 44–56.

14. Bergaust, *Wernher von Braun*, 518–20.

15. Army FOI, 513–17. Basic Personnel Record dated September 20, 1945.

16. von Braun Bibliography.

17. von Braun, "Rocket-Propelled Missile," U.S. Patent no. 2,967,393, issued January 21, 1961.

18. Bergaust, *Wernher von Braun*, 191.

19. Army FOI, 281–89. "Contract for Personal Services (Project Paperclip)" dated July 1, 1951.

20. Bergaust, *Wernher von Braun*, 325–26.

21. "von Braun, Wernher," *Current Biography*, 1952, 607–9.

22. "Space, Here We Come," *Time* September 17, 1951, 58–59.

23. von Braun, *Das Mars Projekt*.

24. von Braun, *The Mars Project*.

25. Army FOI, 576–78. Agent report signed by Richard C. Puglisi, April 21, 1953.

26. Bergaust, *Wernher von Braun*, 156.

CHAPTER 6: SELLING SPACE: THE *COLLIER'S* ARTICLES

1. Stuhlinger and Ordway, *Biographical Memoir*, 133.

2. Whipple, "Recollections of Pre-Sputnik Days," in *Blueprint for Space*, ed. Frederick I. Ordway III and Randy Liebermann, 127–34.

3. Ley, *Rockets* 148.

4. Stuhlinger and Ordway, *Biographical Memoir*, 112.

5. Liebermann, "The *Collier's* and Disney Series," in *Blueprint for Space*, Frederick I. Ordway III and Randy Liebermann, eds., 135–46.

6. Stuhlinger and Ordway, *Biographical Memoir*, 115.

7. von Braun Papers, container 42. Letter of Agreement between *Collier's* and Wernher von Braun dated December 11, 1951.

8. Army FOI, 281–88. Von Braun's contract with the Army dated July 1, 1951.

9. von Braun Papers, container 42. Letter from Willy Ley to von Braun dated February 9, 1952.

10. Ibid., Memo from Seth H. Moseley II, dated March 12, 1952.

11. *Collier's*, March 22, 1952, cover.

12. Ibid., "What Are We Waiting For?" 23.

13. von Braun, Wernher, "Crossing the Last Frontier," *Collier's*, March 22, 1952, 24–29, 72, 74.

14. von Braun Papers, container 46. "General Summary of Professor von Braun's Speech at the Rotary Club," January 16, 1947.

15. von Braun, *The Mars Project*, xvi.

16. von Braun and Ordway, *History*, 202.

17. Ley, Willy. "A Station in Space," *Collier's*, March 22, 1952, 31–32.

18. Whipple, Fred, "The Heavens Open," *Collier's*, March 22, 1952, 32–33.

19. Kaplan, Joseph, "This Side of Infinity," *Collier's*, March 22, 1952, 34.

20. Haber, Heinz, "Can We Survive in Space?" *Collier's*, March 22, 1952, 35, 65–67.

21. Schachter, Oscar, "Who Owns the Universe?" *Collier's*, March 22, 1952, 36, 70–71.

22. "Space Quiz, Around the Editor's Desk," *Collier's*, March 22, 1952, 38–39.

23. *Collier's*, October 18, 1952, cover.

24. von Braun, Wernher, "Man on the Moon: The Journey," *Collier's*, October 18, 1952, 52–60.

25. Eliot, 56–59.

26. "Man on the Moon," *Collier's*, October 18, 1952, 51.

27. von Braun, "The Exploration of the Moon," *Collier's*, October 25, 1952, 38–40.

28. Army FOI, 577–78. Agent report signed Richard C. Puglisi, dated April 21, 1953.

29. von Braun Papers, container 58. Letter to Y. Frank Freeman Jr., from Wernher von Braun dated February 5, 1953.

30. "Picking the Men," *Collier's*, February 28, 1953, 42–48.

31. "Testing the Men," *Collier's*, March 7, 1953, 56–63.

32. "EMERGENCY!" *Collier's*, March 14, 1953, 38–44.

33. von Braun, Wernher, with Cornelius Ryan, "Baby Space Station," *Collier's*, June 27, 1953, 33–35, 38, 40.

34. *Collier's*, April 30, 1954, cover.

35. Whipple, "Is There Life on Mars?" *Collier's*, April 30, 1954, 21.

36. von Braun, Wernher, with Cornelius Ryan, "Can We Get to Mars?" *Collier's*, April 30, 1954, 22–29.

37. von Braun, *The Mars Project*.

38. Ryan, *Across the Space Frontier*.

39. Ryan, *Conquest of the Moon*.

40. Ley, and von Braun, *The Exploration of Mars*.

41. von Braun Resume.

42. Gunston, 37.

43. Ordway and Sharpe, 43.

CHAPTER 7: DISNEYLAND

1. Finch, *Walt Disney's America*, 181.

2. Mosley, 228–30.

3. Ibid., 230–33.

4. Thomas, 267.

5. Mosley, 244.

6. Liebermann, Randy, "The *Collier's* and Disney Series," in *Blueprint for Space*, ed. Frederick I. Ordway III and Randy Liebermann, 144–45.

7. Stuhlinger and Ordway, *Biographical Memoir*, 115–16.

8. Shows, 29–30.

9. Ibid., 31.

10. Wright, Mark, "The Disney-von Braun Collaboration and Its Influence on Space Exploration," www.msfc.nasa.gov/general/history.html

11. Mosley, 265.

12. Liebermann, Randy, "The *Collier's* and Disney Series," in *Blueprint for Space*, ed. Frederick I. Ordway III and Randy Liebermann, 146.

13. Stuhlinger and Ordway, *Biographical Memoir*, 116–17.

14. Koenig, 19, 22–26.

15. Mosley, 226.

16. The author's memories of the original Tomorrowland, Anaheim, Calif.

17. Weiss, Werner, "Welcome to Yesterland," www.msc.net/~werner/yester.html

18. Koenig, 21.

19. Army FOI, 128–130. Memos from C. H. Petrie, January 26, 1954; L. E. Simon, March 25, 1954; H. N. Toftoy, April 13, 1954.

20. Army FOI, 123–27. Statement of Personal History signed by W. von Braun, dated October 15, 1954.

21. Army FOI, 727. Agent report by C. E. Beal, December 3, 1954.

22. "With 102 Others, Von Braun Becomes U.S. Citizen," *Washington Post and Times Herald*, April 15, 1955, 15.

23. Bergaust, *Werner von Braun*, 192.

CHAPTER 8: THE FIRST RACE FOR SPACE

1. von Braun and Ordway, *History*, 159.

2. Ibid., 176–78.

3. Bergaust, *Wernher von Braun*, 231–33.

4. von Braun, Wernher, "Space Superiority," *Ordnance*, 37, no. 197 (March–April 1953): 770–75.

5. Ordway and Sharpe, 375.

6. von Braun and Ordway, *History*, 179.

7. Ordway and Sharpe, 376.

8. Chapman, 101–2.

9. Ibid., p. 111.

10. von Braun and Ordway, *History*, 179.

11. Whipple, Fred L., "Recollections of Pre-Sputnik Days," in *Blueprint for Space*, ed. Frederick I. Ordway III and Randy Liebermann, 127–34.

12. von Braun and Ordway, *History*, 179–80.

13. "With 102 Others, Von Braun Becomes U.S. Citizen," *Washington Post and Times Herald*, April 15, 1955, p. 15.

14. Bergaust, *Wernher von Braun*, photo and caption after 105.

15. "Guided Missiles, Rockets, and Artificial Satellites (Including Project Vanguard)," *Army Library Special Bibliography No. 11*, January 1957 (courtesy of Redstone Arsenal Historical Office).

16. Ordway and Sharpe, 377–78.

17. von Braun Resume.

18. Ordway and Sharpe, 378–79.

19. Gunston, 60.

20. von Braun and Ordway, *History*, 128.

21. Bergaust, *Wernher von Braun*, 243.

22. Medaris, 147.

23. Ordway and Sharpe, 380.

CHAPTER 9: THE SOVIET CHIEF DESIGNER AND SPUTNIK

1. Ordway and Sharpe, 389.

2. Tsiolkovsky, K. E., "K. E. Tsiolkovsky: An Autobiography," *Astronautics*,

May 1959; reprinted in *The Coming of the Space Age*, ed. Arthur C. Clarke, 100–104.

3. von Braun and Ordway, *History*, 40–43.
4. Oberg, 20.
5. Ibid., 17.
6. Vladimirov, 22–23.
7. Ibid., 25–26.
8. Ibid., 28–29.
9. Oberg, 18.
10. Ibid., 18–21.
11. Ibid., 14.
12. Vladimirov, 31.
13. Ibid., 40.
14. McGovern, 159.
15. Ordway and Sharpe, 305–6.
16. Army FOI, 262. Untitled von Braun dossier document.
17. Ordway and Sharpe, 320.
18. McGovern, 206.
19. Ordway and Sharpe, 318–21.
20. Ibid., 46.
21. Ordway and Sharpe, 322.
22. Ibid., 323.
23. McGovern, 214.
24. Ordway and Sharpe, 351.
25. Vladimirov, 43–44.
26. Ordway and Sharpe, 333–34.
27. "The Russian Right Stuff," prod. for *Nova*. (Boston: WGBH for PBS), 1991.
28. Oberg, 24.
29. Ibid., 23.
30. Ordway and Sharpe, 341–42.
31. Army FOI, 145–48. Report of Capt. F. R. Nottrodt to Chief of Ordnance, November 28, 1952.
32. Grey and Grey, 170.
33. Khruschev, 45–47.
34. Gunston, 51.
35. Vladimirov, 48–49.
36. Oberg, 27–28.
37. Vladimirov, 56.
38. Oberg, 30.
39. Vladimirov, 60.
40. von Braun and Ordway, *History*, 164.
41. Oberg, 32.

CHAPTER 10: *EXPLORER I*

1. "Reach for the stars," *Time*, February 17, 1958, 21–25.
2. Medaris, 155.
3. Ordway and Sharpe, 383.

4. von Braun and Ordway, *History*, 162.

5. Shepard and Slayton, 44–45.

6. Medaris, 166–70.

7. Bergaust, *Werner von Braun*, 270.

8. von Braun and Ordway, *History*, 161.

9. Barkdoll, Robert, "Butler Holds von Braun Responsible for Missiles," *Washington Post and Times Herald*, December 7, 1957, A8–A9.

10. Medaris, 186–88.

11. Bergaust, *Reaching for the Stars*, 328–31.

12. Ibid., 319–21.

13. von Braun and Ordway, *History*, 183.

CHAPTER 11: CELEBRITY

1. Joyce, James Avery, "Medal Without Honor," *The Nation*, May 2, 1959, 407–8. Citation by President Dwight D. Eisenhower that accompanied the awarding of a Distinguished Federal Service Medal to Wernher von Braun.

2. "Space on Earth," *Time*, February 17, 1958, 19.

3. "Lyndon at the Launching Pad," *Time*, February 17, 1958, 19–20.

4. "Reach for the Stars," *Time*, February 17, 1958, 21–25.

5. Bergaust, *Reaching for the Stars*, 339–41.

6. von Braun Papers, container 46. Statement to the Elliott Committee on Education and Labor, March 14, 1958.

7. Pryor, Betty, "Oppenheiner 'Waste' Tragic, Says Von Braun," *Washington Post and Times Herald*, March 15, 1958, A2.

8. McDougall, 385–86.

9. von Braun and Ordway, *History*, 164.

10. Shternfeld, 348.

11. Vladimirov, 67–68.

12. Daniloff, 84, 118–19.

13. von Braun, Wernher, "Reminiscences of German Rocketry," *Journal of the British Interplanetary Society*, May 1956; reprinted in *The Coming of the Space Age*, ed. Arthur C. Clarke, pp. 33–55.

14. von Braun, Wernher, "Space Man—The Story of My Life," *The American Weekly*, July 20, 1958, 7–9, 22–25.

15. People, *Time*, September 21, 1959, 46.

16. Ordway and Sharpe, 379.

17. von Braun Resume.

18. von Braun Papers. Letter from Gordon L. Harris to Col. E. W. Richardson, May 16, 1959.

19. Ordway and Sharpe, 390.

20. Army FOI, 273. Letter from DA Washington, DC to CINCUSAREUR Heidelberg, Germany, June 24, 1960.

21. von Braun Papers. Letter from Sen. John Sparkman and Sen. Lister Hill to J. Edgar Hoover, June 6, 1960.

22. For credits, see *I Aim at the Stars*, Dell Movie Classics, 1960. A copy is in von Braun Papers, container 52.

23. Bergaust, *Wernher von Braun*, 368–69.
24. Stuhlinger and Ordway, *Illustrated Memoir*, 52.
25. Cinema, *Time*, October 17, 1960, p. 95.
26. Irving, 295.
27. Ordway and Sharpe, 390.
28. *I Aim at the Stars*, (Dell Movie Classics, 1960). A copy is in von Braun Papers, container 52.

CHAPTER 12: THE CHALLENGE OF THE MOON

1. Kennedy, 244.
2. Aldrin and McConnell, 94.
3. Ordway and Sharpe, 386.
4. Stuhlinger and Ordway, *Biographical Memoir*, 139.
5. Ordway and Sharpe, 387–88.
6. Stuhlinger and Ordway, *Biographical Memoir*, 165.
7. Bergaust, *Wernher von Braun*, 284.
8. "Marshall's Early Years: 1960–1965," www.msfc.nasa.gov/general/history.html
9. "Marshall Center Highlights for 1960," www.msfc.nasa.gov/general/history.html
10. Ibid., xiii–xiv.
11. Ordway and Sharpe, 388. Haeussermann's connection to the team in Germany reported.
12. Army FOI, 97. Memo dated June 3, 1960 from U.S. Army Intelligence Center to MSC Security Officer. subject: Transfer of Investigative Files; Army FOI, 98, memo dated October 10, 1960, from HQ, DA, OACOFS for Intelligence, Washington, to Commanding General U.S. Army Intelligence Center.
13. Bergaust, *Wernher von Braun*, 208.
14. Stuhlinger and Ordway, *Illustrated Memoir*, 60.
15. Schlesinger, 317.
16. Sorensen, 610–13.
17. von Braun and Ordway, *History*, 196.
18. Schlesinger, 233–85.
19. Ibid., 290.
20. Murray and Cox, 80–81.
21. Aldrin and McConnell, 62–63.
22. Schlesinger, 20.
23. Manchester, 1138–39.
24. von Braun and Ordway, *History*, 160, 207.
25. Kennedy, John F., "Special Message to Congress on Urgent National Needs," May 21, 1961; www.cs.umb.edu/jfklibrary
26. Sorensen, 613–17.
27. The author's memory.
28. Wilford, 7–8.
29. Oberg, *Red Star In Orbit*, 39–49.
30. Oberg, *Uncovering Soviet Disasters*, 177–82.

31. Oberg, James E., "Disaster at the Cosmodrome," *Air and Space Smithsonian*, (December 1990–January 1991): 74–77.

32. Oberg, James E., "The Plesetsk Cosmodrome," *Final Frontier*, 36–43.

33. Grey and Grey, 171.

34. von Braun and Ordway, *History*, 170, 172–73.

35. von Braun, Wernher, "Crossing the Last Frontier," *Collier's*, March 22, 1952, 24–29, 72, 74.

36. Murray and Cox, 110.

37. von Braun and Ordway, *History*, 170.

38. Dornberger, 29–30.

39. Ordway and Sharpe, 33.

40. Hunt, Linda, "U.S. Coverup of Nazi Scientists," *Bulletin of the Atomic Scientists* (April 1985): 16–24.

41. "Dr. Kurt H. Debus," www.pao.ksc.nasa.gov/ksc.pao/bios/debus.html

42. von Braun, *First Men to the Moon*, 30–31.

43. Common reference sources.

44. Murray and Cox, 115–20.

45. Aldrin and McConnell, 90.

46. Murray and Cox, 134–35.

47. Ibid., 137–39.

48. Bergaust, *Wernher von Braun*, 409.

49. Wilford, 48.

50. Interview of Edward O. Buckbee in "Wernher von Braun: He Conquered Space," Discovery Channel, 1996.

CHAPTER 13: THE CUBAN MISSILE CRISIS

1. Bergaust, *Wernher von Braun*, 164.

2. Manchester, 1173–76.

3. Brugioni, 201, 210–211.

4. Manchester, pp. 1176–1177.

5. Brugioni, 277.

6. Manchester, 1114–16.

7. Thompson, 268–71.

8. Brugioni, 356–62.

9. Young et al., 8.

10. Thompson, 273.

11. Manchester, 1185–90.

12. Thompson, 345–46.

13. Herken, G., "Waiting for Someone to Blink," *Air and Space Smithsonian*, (Oct.–Nov. 1992): 100–101.

14. Wolfe, 122–25.

15. von Braun and Ordway, *History*, 202–22.

16. Kubrick, Stanley, producer, *Dr. Strangelove* (Columbia Pictures, 1964), Motion picture.

17. Walker, 137.

CHAPTER 14: THE MOONSHIP

1. von Braun, Wernher, "Space Superiority," *Ordnance*, 37 no. 197 (March–April 1953): 770–75.
2. Murray and Cox, 54–55.
3. Collins, 120.
4. Ibid., 122–24.
5. Murray and Cox, 153.
6. Ibid., 160.
7. Bergaust, *Wernher von Braun*, 413–14.
8. Shepard and Slayton, 166.
9. Murray and Cox, 161.
10. Aldrin and McConnell, 107–8.
11. "Marshall Center Highlights for 1966," www.msfc.nasa.gov/general/history.html
12. Wilford, 57.
13. Ibid., 49.
14. Manchester, 1252.
15. Ibid., 1297–98.
16. Ibid., 1298–1300.
17. von Braun Papers. Memoranda and itineraries prepared by Bart J. Slattery Jr., dated April 6 and 7, 1965.
18. Finch, 398.
19. Stuhlinger and Ordway, *Biographical Memoir*, 283–84.
20. Bergaust, *Wernher von Braun*, 370.

CHAPTER 15: BLUNDERS AND DISASTERS

1. von Braun, *First Men to the Moon*, 24.
2. Bergaust, *Wernher von Braun*, 522.
3. von Braun Resume.
4. von Braun Papers, container 3. AAAS membership invitation; Disposition Form signed L. W. Sheeran, Chief Security Office, dated June 18, 1958; letter to AAAS from von Braun dated September 9, 1958.
5. von Braun Bibliography.
6. von Braun, Wernher, "Wernher von Braun Answers Your Questions," *Popular Science*, 182, no. 1 (January 1963): 92.
7. von Braun, Wernher, "More Answers to Your Questions About Space," *Popular Science* 182, no. 2. (February 1963): 92.
8. von Braun, Wernher, "When Will We Land on Mars?" *Popular Science*, 186, no. 3 (March 1965): 86.
9. von Braun, Wernher, "Whatever Happened to the Manned Space Station?" *Popular Science*, 186, no. 4 (April 1965): 87.
10. von Braun and Ordway, *History*.
11. "Marshall Center Highlights for 1970," www.msfc.nasa.gov/general/history.html

12. Stuhlinger and Ordway, *Biographical Memoir*, 256–59.

13. Young et al., 6–8.

14. Wolfe, 122.

15. Fallaci, 404.

16. Ibid., 218–33.

17. FBI FOI. Translation of *The Secret of Huntsville* by Julius Mader and accompanying documents.

18. "Thought Police," *Washington Post*, August 6, 1964, A31.

19. Stuhlinger and Ordway, *Biographical Memoir*, 50–53.

20. Oberg, *Red Star in Orbit*, 87–89.

21. "Launcher Shortage Delays Bion Project," *Science*, 247 (October 25, 1996): 487.

22. Murray and Cox, 201–2.

23. Wilford, 92–101.

24. Ibid., 101–10.

25. Oberg, *Red Star in Orbit*, 91.

26. Oberg, *Uncovering Soviet Disasters*, 171.

27. Ibid., 172–73.

28. Wolfe, 19–43.

29. Common reference sources.

30. Aldrin and McConnell, 193–94, 198.

31. Borman, 195.

CHAPTER 16: MEN ON THE MOON

1. Bergaust, *Wernher von Braun*, 515.

2. Common reference sources.

3. "The Scene at the Cape: Prometheus and a Carnival," *Time*, July 25, 1969, 13.

4. Mailer, 70–71.

5. Ibid., 71–72.

6. Ibid., 72–73.

7. Armstrong et al., 60–61.

8. Ibid., 65–66.

9. Editors' Note, *Life*, August 29, 1969, 1.

10. Mailer, 68.

11. Ibid., 73–74.

12. Ibid., 76–78.

13. Bergaust, *Wernher von Braun*, 420.

14. Armstrong et al., 106–7.

15. Murray and Cox, 101–2.

16. *Time*, July 25, 1969, photo legend.

17. von Braun and Ordway, *Rockets' Red Glare*, 17.

18. "Awe, Hope and Skepticism on Planet Earth," *Time*, July 25, 1969, 16.

19. Aldrin and McConnell, 227.

20. Armstrong et al., 98.

21. *Holy Bible*, Revised Standard Version, Matt. 6: 9–13.

22. Mailer, 100–101.

23. Armstrong et al., pp. 103–104.

24. Aldrin and McConnell, 228–29.

25. Ibid., 233.

26. Ibid., 231.

27. Oberg, *Red Star in Orbit*, 124–26.

28. Armstrong et al., 17–25.

29. Aldrin and McConnell, 234–38.

30. Ibid., 239–45.

31. Wilford, 269–82.

32. Ibid., 283.

33. Ibid., 284.

34. Armstrong et al., 424–25.

35. Stuhlinger and Ordway, *Illustrated Memoir*, 84.

36. von Braun Papers, container 47. "Afternoon Picnic Speech," July 26, 1969.

37. von Braun Papers, container 47. "Lunar Landing Celebration Dinner Speech," July 26, 1969.

38. FBI FOI, letter from J. Edgar Hoover to John D. Ehrlichman dated July 30, 1969.

39. Oberdorfer, Don, "Apollo Astronauts Hailed by Millions," *Washington Post*, August 14, 1969, 1, 8.

40. Thackery, Ted Jr., "Star-Spangled Fete for Moon Pioneers," *Los Angeles Times*, August 14, 1969, part 1, 1, 2.

41. "White House Lists Guests at Astrofete," *Los Angeles Times*, August 14, 1969, Part 4, 1, 7–9, 11, 14, 20.

CHAPTER 17: SHUTTLE, SPACE STATION, AND THE DECLINE OF NASA

1. von Braun, Wernher. "Man on the Moon," *Los Angeles Times*, (Space/66), June 19, 1966, 8, 16, 31.

2. Hunt, *Secret Agenda*, 226, 311.

3. USA v. Andrae.

4. FBI FOI. Letter from J. Edgar Hoover to John D. Ehrlichman dated July 30, 1969.

5. FBI FOI. Memo to Director from FBI Philadelphia dated September 16, 1996.

6. von Braun, Wernher (Introduction by Ron Miller), "Now That Man Has Reached the Moon?", in Ordway & Liebermann, pp. 167–75.

7. McConnell, 31.

8. Trento, 91.

9. *Report of the Presidential Commission*, 2.

10. Wilford, xviii, 67.

11. McDougall, 383.

12. Wilford, 128.

13. Buckbee, Edward O., and Walker, Charles, "Spaceflight and the Public Mind," in *Blueprint for Space*, ed. Frederick I. Ordway III and Randy Liebermann, 189–97.

14. Stuhlinger and Ordway, *Biographical Memoir*, 284–85.

15. "Background Story," Huntsville, Ala.: U.S. Space & Rocket Center.

16. Finch, 398.

17. Weiss, Werner. "Welcome to Yesterland," www.msc.net/˜werner/yester.html

18. Trento, 89–90.

19. Stuhlinger and Ordway, *Biographical Memoir*, 291.

20. Ordway and Sharpe, 401.

21. Trento, 94.

22. Ordway and Sharpe, 403.

23. FBI FOI, unsigned letter to Alexander P. Butterfield dated November 5, 1970.

24. Ibid., FBI report dated October 5, 1970.

25. Ibid., Report of FBI interview of von Braun dated November 12, 1970.

26. Collins, 163–64.

27. "The Historical Threads of Space Station: Early Space Station Concepts," www.msfc.nasa.gov/general/history.html

28. Trento, 93.

29. *Report of the Presidential Commission*, 3.

30. Stockton and Wilford, 35–37 and photo between 88 and 89.

31. Trento, 107.

32. "Emergency," *Collier's*, March 14, 1953, 38–44.

33. Stockton and Wilford, 42.

34. Hunt, *Secret Agenda*, 228–29.

35. Ordway and Sharpe, 397.

36. Hunt, *Secret Agenda*, 225.

37. "Dr. Kurt H. Debus," NASA Biography online, www.pao.ksc.nasa.gov/ksc.pao/bios/debus.htm

38. "Marshall Center Highlights for 1973," www.msfc.nasa.gov/general/history.html

39. Bergaust, 194.

40. Hunt, *Secret Agenda*, 229.

41. Bergaust, *Wernher von Braun*, 461.

42. Bernstein and Woodward, 13–16.

CHAPTER 18: EXILE IN AMERICA: VON BRAUN'S LAST YEARS

1. Ordway and Sharpe, 404.

2. Stuhlinger and Ordway, *Biographical Memoir*, 309.

3. Bergaust, *Wernher von Braun*, 460.

4. Ibid., 463–66.

5. Stuhlinger and Ordway, *Biographical Memoir*, 309–11.

6. Ibid., 325.

7. Bergaust, *Wernher von Braun*, 469.

8. Ibid., 521.

9. Bergaust, *Wernher von Braun*, 107.

10. Ibid., 467–69.

11. Stuhlinger and Ordway, *Biographical Memoir*, 312–13.
12. Collins, 12.
13. Ordway and Sharpe, 397.
14. Collins, 274.
15. Ordway and Sharpe, 318.
16. Bergaust, *Wernher von Braun*, 527–29.
17. Stuhlinger and Ordway, *Biographical Memoir*, 98.
18. Ibid., 118.
19. Ibid., 217.
20. Collins, 159.
21. Bergaust, *Wernher von Braun*, 526.
22. Ibid., 544–45.
23. Stuhlinger and Ordway, *Biographical Memoir*, 328.
24. Ordway and Sharpe, 404.
25. Stuhlinger and Ordway, *Biographical Memoir*, 328–29.
26. Ibid., 326–27.

CHAPTER 19: EPILOGUE: THE MAN WHO SOLD THE MOON

1. Stuhlinger and Ordway, *Biographical Memoir*, 250.
2. McGovern, 120–22.
3. Stuhlinger and Ordway, *Biographical Memoir*, 51–53.
4. Lindbergh, 991–98.
5. Hunt, *Secret Agenda*, 230–31.
6. Sherrill, Robert, "The Golden Years of An Ex-Nazi," *The Nation* (June 7, 1986): cover; 792–96.
7. Michel, 1979.
8. Ibid., 2–3.
9. Ibid., 89.
10. Speer, *Infiltration*, 210–11.
11. Irving, 204. The author noted that von Braun visited the Central Works (Mittelwerk) on January 25, 1944.
12. Michel, 90–92.
13. Hunt, *Secret Agenda*, 239–240.
14. USA v. Andrae.
15. Hunt, *Secret Agenda*, 240–41.
16. Ibid., 245–46.
17. Faludi, Susan, "In the Rocket's Glare," *Atlanta Weekly (The Atlanta Journal, The Atlanta Constitution)*, May 26, 1985.
18. Blumenthal, Ralph, "German-Born NASA Expert Quits U.S. to Avoid a War Crimes Suit," *New York Times* October 18, 1985, A1, A5.
19. Hunt, *Secret Agenda*, 254.
20. Army FOI.
21. FBI FOI.
22. O'Toole, Thomas, and Mary Thornton. "Road to Departure of Ex-Nazi Engineer," *The Washington Post* November 4, 1984, A1, A25.

23. Hunt, *Secret Agenda*, 65.

24. Neufeld, 228.

25. Saxon, Wolfgang, "NASA Hero Arthur Rudolph Dies in Native Germany at 89," *San Jose Mercury News* January 3, 1996, 4A.

26. Fisher, Marc, "Germany Cancels V-2 Launch Fete," *The Washington Post* September 29, 1992, A17.

Bibliography

All books used as references are listed in this bibliography. Newspaper and periodical articles as well as Internet sources are given fully in the notes, and are not repeated here.

Agoston, Tom. *Blunder: How the U.S. Gave Away Nazi Supersecrets to Russia.* New York: Dodd, Mead and Company, 1985.

Aldrin, Buzz, and Malcolm McConnell. *Men from Earth.* New York: Bantam Books, 1989.

Armstrong, Neil, Michael Collins, and Edwin E. Aldrin Jr. *First on the Moon.* Boston, Mass.: Little, Brown and Company, 1970.

Bar-Zorah, Michel. *The Hunt for German Scientists.* New York: Hawthorn Books, 1967.

Bergaust, Erik. *Reaching for the Stars.* Garden City, N.Y.: Doubleday and Company, 1960.

———. *Wernher von Braun.* Washington, D.C.: National Space Institute, 1976.

Bernstein, Carl, and Bob Woodward. *All the President's Men.* New York: Touchstone, Simon & Schuster, 1974.

Borman, Frank. *Countdown.* New York: Silver Arrow Books, William Morrow, 1988.

Bowe, Tom. *The Paperclip Conspiracy.* Boston: Little, Brown and Company, 1987.

Brugioni, Dino A. *Eyeball to Eyeball: The Inside Story of the Cuban Missile Crisis.* New York: Random House, 1991.

Churchill, Winston S. *The Second World War: Triumph and Tragedy.* Vol. 6. Boston: Houghton Mifflin Company, 1953.

Clarke, Arthur C., ed. *The Coming of the Space Age.* New York: Meredith Press, 1967.

Collins, Michael. *Liftoff: The Story of America's Adventure in Space.* New York: Grove Press, 1988.

David, Heather M. *Wernher von Braun.* New York: Putnam, 1967.

Dornberger, Walter. *V-2.* New York: Viking Press, 1954.

Eliot, T. S. *The Complete Poems and Plays 1909–1950.* New York: Harcourt, Brace and World, 1971.

Ethell, Jeffrey L. *Komet: The Messerschmitt 163.* New York: Sky Books Press, 1978.

Fallaci, Oriana. *If the Sun Dies.* New York: Atheneum, 1966.

Finch, Christopher. *The Art of Walt Disney.* New York: Harry N. Abrams, 1973.

———. *Walt Disney's America.* New York: Abbeville Press, 1978.

Garlinski, Josef. *Hitler's Last Weapons.* New York: Times Books, 1978.

Grey, Jerry, and Vivian Grey. *Space Flight Report to the Nation.* New York: Basic Books, 1962.

"Guided Missiles, Rockets, and Artificial Satellites (Including Project Vanguard)." Army Library Special Bibliography No. 11. January 1957. Redstone Arsenal Historical Office.

Gunston, Bill. *The Illustrated Encyclopedia of the World's Rockets and Missiles.* New York: Crescent Books, 1978.

Hilberg, Raul. *The Destruction of the European Jews.* Chicago: Quadrangle Books, 1961.

Hoess, Rudolf. *Commandant of Auschwitz.* New York: The World Publishing Company, 1959.

Hunt, Linda. *Secret Agenda: The United States Government, Nazi Scientists, and Project Paperclip, 1945 to 1990.* New York: St. Martin's Press, 1991.

Huzel, Dieter K. *Peenemuende to Canaveral.* Englewood Cliffs, N.J.: Prentice-Hall, 1962.

Irving, David. *The Mare's Nest.* Boston, Mass.: Little, Brown and Company, 1964.

Joyce, James Avery. "Medal Without Honor." *The Nation* (May 2, 1959), 407–8.

Kennedy, John F. *The Burden and the Glory,* edited by Allan Nevins. New York: Harper and Row, 1964.

Khrushchev, Nikita. *Khrushchev Remembers: The Last Testament.* Translated and edited by Strobe Talbot. Boston, Mass.: Little, Brown and Company, 1974.

Koenig, David. *Mouse Tales: A Behind the Ears Look at Disneyland.* Irvine, Calif.: Bonaventure Press, 1994.

Krausnick, H., H. Bucheim, M. Broszat, and H.-A. Jacobsen. *Anatomy of the SS State.* New York: Walker and Company, 1965.

Kubrik, Stanley. *Dr. Strangelove.* Hollywood: Columbia Pictures, 1964.

Ley, Willy. *The Conquest of Space.* New York: Viking Press, 1949.

———. *Rockets, Missiles, and Men in Space.* New York: Viking Press, 1968.

Ley, Willy, and Wernher von Braun. *The Exploration of Mars.* New York: Viking Press, 1956.

Lindbergh, Charles A. *The Wartime Journals of Charles A. Lindbergh.* New York: Harcourt Brace Jovanovich, 1970.

Lomax, Judy. *Flying for the Fatherland: The Century's Greatest Pilot.* New York: Bantam Books, [paperback], 1991.

Mailer, Norman. *Of a Fire on the Moon.* Boston, Mass.: Little, Brown and Company, 1970.

Manchester, William. *The Glory and the Dream: A Narrative History of America 1932–1972.* Vols. 1 and 2. Boston, Mass.: Little, Brown and Company, 1974.

McConnell, Malcolm. *Challenger: A Major Malfunction.* Garden City, N.Y.: Doubleday and Company, 1987.

McDougall, Walter A. *The Heavens and the Earth.* New York: Basic Books, 1985.

McGovern, James. *Crossbow and Overcast.* New York: William Morrow and Co., 1964.

Medaris, John B. *Countdown for Decision.* New York: G. P. Putnam's Sons, 1960.

Michel, Jean. *Dora.* New York: Holt, Rinehart and Winston, 1979.

Mosley, Leonard. *Disney's World: A Biography.* Briarcliff Manor, N.Y.: Stein and Day, 1985.

Murray, Charles, and Catherine Bly Cox. *The Race to the Moon.* New York: Simon and Schuster, 1989.

Neufeld, Michael J. *The Rocket and the Reich: Peenemuende and the Coming of the Missile Era.* New York: The Free Press, 1995.

Oberg, James E. *Uncovering Soviet Disasters: Exploring the Limits of Glasnost.* New York: Random House, 1988.

————. *Red Star in Orbit.* New York: Random House, 1981.

Office of the United States Chief of Counsel for Prosecution of Axis Criminality. *Nazi Conspiracy and Aggression.* Vol. 6. Washington, D.C.: United States Government Printing Office, 1946.

Ordway, Frederick I. III, and Randy Liebermann, eds. *Blueprint for Space.* Washington, D.C.: Smithsonian Institution Press, 1992.

Ordway, Frederick I. III, and Mitchell R. Sharpe. *The Rocket Team.* New York: Thomas Y. Crowell, 1979.

Persico, Joseph E. *Piercing the Reich.* New York: Ballantine Books [paperback], 1979.

Piszkiewicz, Dennis. *The Nazi Rocketeers: Dreams of Space and Crimes of War.* New York: Praeger, 1995.

Reitsch, Hanna. *Flying Is My Life.* New York: G. P. Putnam's Sons, 1954.

Report of the Presidential Commission on the Space Shuttle Challenger Accident. Washington, D.C.: U.S. Government Printing Office, 1986.

Ryan, Cornelius, ed. *Across the Space Frontier.* New York: Viking Press, 1952.

————. *Conquest of the Moon.* New York: Viking Press, 1953.

Schlesinger, Arthur M. Jr. *A Thousand Days: John F. Kennedy in the White House.* New York: Greenwich House, 1965.

Shepard, Alan, and Deke Slayton. *Moon Shot.* Atlanta, Ga.: Turner Publishing, 1994.

Shirer, William L. *The Rise and Fall of the Third Reich.* New York: Simon and Schuster, 1960.

Shows, Charles. *Walt: Backstage Adventures with Walt Disney.* La Jolla, Calif.: Windsong Books International, 1979.

Shternfeld, Ari. *Soviet Space Science.* 2nd ed. New York: Basic Books, 1959.

Sorensen, Theodore C. *Kennedy*. New York: Harper and Row, 1965.

Speer, Albert. *Infiltration*. New York: Macmillan Publishing Co., 1981.

———. *Inside the Third Reich*. New York: The Macmillan Company, 1970.

Stockton, William, and John Noble Wilford. *Spaceliner*. New York: Times Books, 1981.

Stuhlinger, Ernst, and Frederick I. Ordway III. *Wernher von Braun, Crusader for Space: A Biographical Memoir*. Malbara, Fla.: Krieger Publishing Company, 1994.

———. *Wernher von Braun, Crusader for Space: An Illustrated Memoir*. Malbara, Fla.: Krieger Publishing Company, 1994.

Thomas, Bob. *Walt Disney: An American Original*. New York: Pocket Books, 1976.

Thompson, Robert Smith. *The Missiles of October: The Declassified Story of John F. Kennedy and the Cuban Missile Crisis*. New York: Simon and Schuster, 1992.

Trento, Joseph J. *Prescription for Disaster*. New York: Crown Publishers, 1987.

Tsiolkovsky. K. E. "K. E. Tsiolkovsky: An Autobiography." In *The Coming of the Space Age*, edited by Arthur C. Clarke, 100–104. New York: Meredith Press, 1967.

Vladimirov, Leonid. *The Russian Space Bluff*. New York: The Dial Press, 1973.

von Braun, Wernher. *First Men to the Moon*. New York: Holt, Rinehart and Winston, 1960.

———. *Das Mars Projekt*. Essling, Germany: Bechtle Verlag, 1952.

———. "Konstructiv, theoretische und experimentalle Beitraege zu dem Problem der Fluessigkeitsraket." Ph.D. diss., Der Friedrich-Wilhems-Universitat zu Berlin, 1934.

———. *The Mars Project*. Urbana: University of Illinois Press, 1953; 1991 edition.

———. "Reminiscences of German Rocketry." *Journal of the British Interplanetary Society* (May 1956). Reprinted in Arthur C. Clarke, ed., *The Coming of the Space Age*, 33–55. New York: Meredith Press, 1967.

———. "Rocket-Propelled Missile." U.S. Patent No. 2,967,393. Issued January 21, 1961.

von Braun, Wernher, and Frederick I. Ordway III. *History of Rocketry and Space Travel*. New York: Thomas Y. Crowell Company, 1966.

———. *The Rockets' Red Glare*. Garden City, N.Y.: Anchor Press/Doubleday, 1976.

Walker, Alexander. *Peter Sellers: The Authorized Biography*. New York: Macmillan Publishing Company, 1981.

Wilford, John Noble. *We Reach the Moon: The New York Times Story of Man's Greatest Adventure*. New York: Bantam Books, 1969.

Wolfe, Tom. *The Right Stuff*. New York: Farrar, Straus and Giroux, 1979.

Wistrich, Robert. *Who's Who in Nazi Germany*. New York: Macmillan Publishing Company, 1982.

Young, Hugo, Bryan Silcock, and Peter Dunn. *Journey to Tranquility*. Garden City, N.Y.: Doubleday and Company, 1970.

Index

About the Author

DENNIS PISZKIEWICZ has been an enthusiast of space exploration since his childhood in the 1950s. He has taught college-level chemistry and biochemistry and has been the recipient of a NASA fellowship. His interest in the history of science and technology inspired him to write *The Nazi Rocketeers: Dreams of Space and Crimes of War* (1995) and *From Nazi Test Pilot to Hitler's Bunker: The Fantastic Flights of Hanna Reitsch* (1997), both published by Praeger.